T0272223

INSIDE SCIENCE

REVOLUTION IN BIOLOGY
AND ITS IMPACT

ALSO FROM COLD SPRING HARBOR LABORATORY PRESS

Inside the Orphan Drug Revolution: The Promise of Patient-Centered Biotechnology

A Cure Within: Scientists Unleashing the Immune System to Cure Cancer

Orphan: The Quest to Save Children with Rare Genetic Disorders

The Strongest Boy in the Word: How Genetic Information Is Reshaping Our Lives, Updated and Expanded Edition

Is It in Your Genes? The Influence of Genes on Common Disorders and Diseases that Affect You and Your Family

Abraham Lincoln's DNA and Other Adventures in Genetics

Conscience and Courage: How Visionary CEO Henri Termeer Built a Biotech Giant and Pioneered the Rare Disease Industry

INSIDE SCIENCE

REVOLUTION IN BIOLOGY
AND ITS IMPACT

Benjamin Lewin

COLD SPRING HARBOR LABORATORY PRESS
Cold Spring Harbor, New York • www.cshlpress.org

INSIDE SCIENCE
Revolution in Biology and Its Impact

Published by Cold Spring Harbor Laboratory Press, Cold Spring Harbor, New York
Printed in the United States of America

Publisher and Acquisitions Editor	John Inglis
Project Manager	Inez Sialiano
Permissions Coordinator	Carol Brown
Production Editor	Kathleen Bubbeo
Production Manager and Cover Designer	Denise Weiss

Front cover image: Chris Knapton/Photodisc via Getty Images.

Library of Congress Cataloging-in-Publication Data

Library of Congress Control Number: 20239003921 | ISBN 978-1-621825-01-2 (hardcover)
| ISBN 978-1-621825-02-9 (ePub3)

10 9 8 7 6 5 4 3 2 1

All World Wide Web addresses are accurate to the best of our knowledge at the time of printing.

For a complete catalog of all Cold Spring Harbor Laboratory Press publications, visit our website at www.cshlpress.org

In the hope that Emma, Stella,
James, Daniel, and Noah
will grow up in a more rational world.

CONTENTS

PART 4: HISTORY

PART 5: LANDMARKS

EPILOGUE

PREFACE

Perhaps this is the time to reveal how I became a critic of science instead of a researcher. Working on an experiment one afternoon as a graduate student, I noticed that everyone had left the lab. I discovered them in a room at the far end of the building. When I asked why they were there, they explained that the procedure I was following could lead to producing the explosive TNT by a side-reaction, so they thought it safest to evacuate. This lack of confidence in my practical prowess gave me some pause for thought.

Unlike the humanities, science has never had a tradition of divorcing criticism from research. The assumption is that scientists do research; indeed, there is a view that this is the *only* legitimate role for a scientist. When I became Editor of *Nature New Biology* in 1971, *Nature* was one of the very few scientific journals in the world where decisions on publication were taken by full-time editors. The Editor of almost every other scientific journal was a researcher in the field; editing a journal was a part-time activity. It was not surprising, therefore, that when I started *Cell* in 1974, I encountered some skepticism.

Aside from the issue of practical prowess, my problem in following a career in research was that it seemed so narrow. You got to know absolutely everything about one aspect of science, but often at the price of being unable to see the broad picture. The increasing specialization of science since then has only exacerbated the problem. I wanted to understand broad swathes of science.

Modern society has been shaped by tremendous advances in science and technology. These hold enormous promise, but also dangers: to assess this rationally requires some understanding of what constitutes science and its limitations. During the COVID pandemic, there were frequent cries of "follow the science," but failure to understand the assumptions underlying the science, and therefore its limitations, may have been responsible for some of the difficulties in dealing with the pandemic.

The value system of science is nonpareil: self-contained and self-validating. In this book, I want to explain how science works. This is not an idealized view; it is science, warts and all. I try to show not only how science should, and often does, work, but also how failings in the system can misdirect it.

This sounds as though I want to dispel the mystique of science. Well, yes and no. I believe fervently in the distinctive, in fact the unique, value system of science. But it does have its flaws. Some are imposed by the institutional framework within which it functions—especially the means of communication (publishing results as research articles), and funding (the need to apply for grants on a continuing basis). Some are due to failings on the part of its practitioners (especially excessive conformance to conventional wisdom). I believe in any case that science benefits from being seen clear-eyed.

The history and philosophy of science take what you might call a "classical" view: that science is practiced by individuals who obtain data to test hypotheses. The main questions I want to ask are how far this description was true, and how has the basis of the scientific endeavor changed in this century.

The trend today marks a move away from individual investigation. "Big science" uses massive amounts of data to replace individual experiments, and entails a different way of thinking about science. A move to use AI (artificial intelligence) techniques to analyze data even raises the question of how long science will continue to be driven by human intellect.

The effects of these changes outside science have scarcely been noticed. Science is regarded as something of a black box: so long as it delivers the goods—whether in the form of better medicine or technological spin-offs that improve daily life—society is prepared to pay the bill without too much concern as to just how those results are achieved. Yet if there is to be informed consent to the progress of science, it is necessary to understand its nature, and the implications of the way it is changing. The human impact of science to date is undeniable, but we have to think about how that may be enhanced by the changes in science itself.

I was concerned when I started to write this book that science might be losing its way. Were scientists being replaced by technicians? Now I believe the situation is not so drastic, although certainly things have changed. My purpose here is to consider the consequences of that change. The main concern is not so much with how information is obtained, but with the capacity to make intelligent use of it.

I start with the major change in science of the past two or three decades: the move from small-scale research in groups of no more than a few people, led by one principal investigator, to a larger scale requiring many researchers, directed in a top-down manner, sometimes by a committee rather than an individual. What does this mean for the very nature of science as well as for the activities of the participants? Has science—in particular biology—changed from testing hypotheses to trawling for data?

Then I look at the medical and industrial implications of molecular biology, before turning to how scientists communicate, how research is funded, and how science is impacted by politics and ethics. All through this, I try to show how the value system of science leads to validation of scientific results, what its limitations may be, and how research reports should be assessed.

I take my illustrations of the traditional scientific process from what many regard as the golden age of molecular biology, a unique period in biological science, because the discoveries were so fundamental and continuous, creating an unparalleled intellectual furor. I suppose it lasted from around 1960 for more than a quarter century. I draw on this period because the sheer pace of discoveries magnifies the sense of what science is about, how it works, and (sometimes) how it doesn't work.

The principle would be the same whether you consider a century of chemistry or physics as opposed to a quarter century of molecular biology, but the focus is sharper. Comparing the history of this period with the subsequent quarter century allows us to ask what we have lost or gained by the change in the way science has been conducted since the golden age.

For a view of what leads to (or impedes) great discoveries, I follow the history of changing views of the gene and DNA, from Mendel, through Watson and Crick, to the present day. Later I look at epigenetics, which gives alternative views of the working of heredity. The rise and fall of dogmas illustrate the role of fashion in science, with the jury still out on epigenetics. Finally it is time to consider the aspects of biological science that most directly impact human life: the sequencing of the human genome and the development of gene editing and its potential.

I draw on my experience at *Cell* to explain how science really works. *Cell* became an important journal during the golden age, publishing many of the most significant papers in biology. The submission, review, and reaction to those papers, especially from behind the scenes, reveals a good deal about the conduct of science. I look at how the trends from that period have accelerated into the present. I want to explain what happens when you "follow the data," and why sometimes the structure of science prevents that from happening.

After years immersed in science, I spent a period (not as long) immersed in something else—the world of wine, as a matter of fact. With the objectivity of greater distance, some of my implicit assumptions have become clearer, and I have changed my views on some issues of scientific conduct. The question today is whether and how science can continue to deliver the goods, both as intellectual stimulus for its participants, and in the form of benefits for the population.

Benjamin Lewin
October 2022

ACKNOWLEDGMENTS

One of the pleasures in writing this book has been the resumption of discussions with scientists with whom I used to interact frequently as Editor of *Cell*.

People who were kind enough to share their recollections or correspond or provide materials include David Baltimore, Adrian Bird, Tom Caskey, Tom Cech, Caroline Dean, Bob Gallo, Demis Hassabis, Lee Hood, Francisco Mojica, Svante Pääbo, Thoru Pederson, Stan Prusiner, Mark Ptashne, Rich Roberts, Phil Sharp, Azim Surani, Harold Varmus, and Jan Witkowski.

I am enormously indebted to colleagues and friends who read chapters, or even the entire book, to suggest many helpful changes, including Caroline Dean, Jamie Kass, Anna Lau, Leslie Pond, Leslie Ann Roldan, Andrea Smith, Ian Trackman, Larry Walker, and Paul Wassarman.

It is especially fitting that this book should be published by the Cold Spring Harbor Laboratory Press, as the idea was conceived during a meeting at Cold Spring Harbor. It's a pleasure to thank John Inglis and the team at Cold Spring Harbor Laboratory Press for the enthusiasm and professionalism with which they brought this project to fruition.

INTRODUCTION

WHAT IS SCIENCE?

At one pole, the scientific culture really is a culture, not only in an intellectual but also in an anthropological sense… . Biologists more often than not will have a pretty hazy idea of contemporary physics; but there are common attitudes, common standards and patterns of behavior, common approaches and assumptions.[1]

C.P. Snow, 1959

Philosophers cannot insulate themselves against science. Not only has it enlarged and transformed our vision of life and the universe enormously; it has also revolutionized the rules by which the intellect operates.[2]

Claude Lévi-Strauss, 1991

In the span of one human lifetime, we have gone from ignorance about the nature of the genetic material to knowledge of how to edit any gene within it. In a previous life span, we went from defining the atom to splitting the atom, and entered the nuclear age. If the twentieth century was the era of physics, the twenty-first century will be the era of biology.

The idea for this book came from a meeting held at Cold Spring Harbor Laboratory on Long Island in New York in 2017 to celebrate the 100th anniversary of Francis Crick's birth. Listening to the talks, many of which focused on work done around and after the time of the discovery of the double helix in 1953, I was struck by the great difference in the way science was done then from how it is practiced now.

As Editor of *Cell* from 1974 to 1999, I saw a significant change in the atmosphere of science, with a transition from research performed principally by individual researchers to working in large groups with a team leader, from testing hypotheses to trawling for data, and from regarding science as an abstruse intellectual pursuit to viewing it with reference to its relevance to modern life. The question in my mind is how far these transitions have changed the very nature of science.

Science is founded on a unique proposition: that it is a self-verifying system in which everyone can play both sides of the game, as either researcher or reviewer. The two sides come together in the key unit of science, the research paper. This is the basic means by which scientists communicate with each other. Researchers write a paper when they feel they have enough data or new ideas to draw interesting conclusions. They submit the paper to a journal, which sends it out for review to two or more other researchers who are experts in the area. This is the much-vaunted system of peer review. On another occasion, roles might be reversed, and the authors of this paper might be reviewing one submitted by the experts who reviewed it. We'll come later to the questions of conflict of interest that can arise from playing dual roles.

The crucial feature of a research paper is that it should have enough information to enable others to reproduce the findings. Other researchers will probably not set out to reproduce a published paper directly, but they may well perform experiments based on its conclusions. If their experiments are inconsistent with the published work, sooner or later there will be experiments to repeat it, and this will lead ultimately to one set of experiments being accepted and one rejected. This is the self-correcting mechanism of science. Arguments about which reality is right are a standard part of the give and take of science.

A research paper is never an end: it is a means to the next paper. It may test a specific hypothesis or report the collection of a set of data, but it is in effect part of a work in progress, subject to reassessment as research continues. One problem with a wider public acceptance of science is that it requires some understanding of the subject and the techniques used in research. The need to master jargon and specialized techniques impedes a wider understanding.

Sociologist Robert Merton made a classic analysis of the value system of science in the 1940s, when he identified four features: communal activity builds on previous efforts (communality); results are independent of whoever makes the discovery (universalism); science is impartial with respect to whatever results emerge (disinterestedness); and science is subject to testing (organized skepticism). "Because of the practice of this ethos, the activity of the scientists is so productive and so different from the babbling and agitating of the ideologues and of the politicians," Merton said.[3]

Although this perhaps somewhat idealistic formulation has been attacked by later sociologists[4]—and it is, of course, the antithesis of post-modernist thinking—it captures the essence of why science represents a unique value system: it is based on facts that are tested objectively by each subsequent contribution. Applying this description, mathematics as well as physics, chemistry, and biology would qualify as science: social sciences are more questionable given the difficulties in establishing controls and verification.[5] Systems that appear to be scientific, but that are not based on hypotheses that can be tested, are (pejoratively) called pseudoscience.

This formulation is a fair account of the results of applying the scientific approach, but it does not take full account of, perhaps it even ignores, the human element. Aside from Merton, anthropologists may have largely, in fact almost entirely, misunderstood the functioning of science, but they are right in the principle that it is more than a methodology: it is a culture. Scientists take positions that are not always driven by objective views of the data; there are competition, ambition, petty rivalries, and from time to time even dishonesty. But science rises above these deficiencies because ultimately its self-correcting character means that data must triumph. Of course, science is a human endeavor, but the human ability to (mis)interpret the data is limited by the nature of the enterprise.

Each branch of science is admittedly a specialized discipline. It is not necessarily easy for scientists in one branch of specialization to understand the details of work in other branches. Making science intelligible to the nonscientist is a specialized task in itself. But my point is that understanding science, at any level beyond simply accepting its conclusions sight unseen, requires more than comprehending the details: it requires coming to grips with the scientific attitude, and understanding the limitations of the scientific process. Failure to do this results in failure to distinguish between science and pseudoscience.

Perhaps because the methodology of science is its own world, it is poorly understood by nonscientists. When he defined the difference between science and the humanities as *The Two Cultures* in the Rede Lecture at Cambridge University in 1959, C.P. Snow created a furor. He was ferociously attacked by the literary establishment for arguing that science was a driving intellectual force in society.

His most famous criticism was that so-called intellectuals are, in fact, illiterate about science. "I have been present at gatherings of people who ... are thought highly educated and who have with considerable gusto been expressing their incredulity at the illiteracy of scientists.... . I have asked the company how many of them could describe the Second Law of Thermodynamics. The response was cold: it was also negative. Yet I was asking something which is about the scientific equivalent of: have you read a work of Shakespeare's?"[6]

This quotation came to epitomize the issue, to the extent that it was caricatured in a song in a musical performance on the London stage, which began, "The first law of thermodynamics, Heat is work and work is heat."[7] Snow reinforced his position by saying later in his lecture, "Intellectuals, in particular literary intellectuals, are natural Luddites."[8]

Perhaps the sharpest difference between science and the humanities is that nothing, but nothing, is immune from questioning in science, whereas outside of science, especially in what used to be regarded as traditional classical education, there were "givens" that could not be questioned. There had been little change in the traditional attitude since Matthew Arnold argued in an earlier Rede Lecture, in 1882, that "no one could be really educated unless they understood literature, particularly the literature of ancient Greece and Rome." Certainly, there are "givens" in science, no better epitomized than in the Second Law of Thermodynamics, but they stand on a basis of objective data that can, in fact, be questioned at any time. There is no equivalent in the humanities, especially in the subjectivity of literary criticism—the source of the most vicious criticism of *The Two Cultures*.

Snow thought the problem of the two cultures was a consequence of the English educational system. Not only was there a gulf between science and the humanities, but it extended to a hierarchy within science. Certainly, when I went through the British school system, the brightest pupils were directed to Latin and Greek, the next group did physics and chemistry, and the bottom group did biology. This was already out of kilter with intellectual reality, and could scarcely have been a greater misreading of developments to come in the rest of the century. The Liberal Arts curriculum in American universities showed something of the same disdain for science, if not so stratified.

It is apparent today that incomprehension of science is a world-wide problem, but it does seem to be strongest in the Anglo-Saxon world, and perhaps at its peak in England. In typical English fashion, the educational system's response to C.P. Snow was not to educate arts graduates about science, but to take the view that supposedly illiterate scientists should be taught to write English, more or less the inverse of what was really needed. The situation has improved little in the following half-century.

Even today, on both sides of the Atlantic, although there is nominally greater acceptance of the idea that you cannot be truly educated unless you understand science, there is still a substantial proportion of the population for whom the concepts of science are impenetrable. (I suppose what seems to be an increasing number who are suspicious of science, or who utterly reject it, would be at the far extreme of that proportion.) When my first book on wine was nominated for the André Simon Award, at the award ceremony, the Chair of the committee described it by saying, "and it has graphs and things in it," with a tone of rising incredulity in her voice approaching a note of horror. I knew at once that I would not be awarded any prize. Yet surely innumeracy should be regarded as just as unacceptable as illiteracy.

I would argue that in the half-century since Snow distinguished between the two cultures, the gap has if anything widened. There remain basic differences in comprehension, with the humanities failing to appreciate the sciences and vice versa, but, more to the point, there is a fundamental difference in attitudes toward knowledge. The humanities can be attractive because they admit a turbulence of ideas, with room to perpetuate completely contrasting views. But although there can, of course, be contrasting interpretations in science, and indeed arguments about the legitimacy of data, ultimately these give way to what scientists call objectivity, and those in the humanities might call the tyranny of data. That difference in attitude—the acceptance that ideas are subservient to facts—versus the view that all ideas can be legitimate, is a basic difference between scientists and nonscientists. A prime objective of this book is to explore how far that attitude is justified in science.

The "scientific attitude," searching for objective truth, is the universal ideal of science. Whether physics, chemistry, or biology, science proceeds by obtaining data through controlled experiments that can be reproduced or challenged by others. However, biology is different in some important respects from physics or chemistry. Living organisms have an intrinsic variability that is different from the invariance of inorganic material. Molecular biology is the part of biology closest to physics or chemistry; experiments are performed in the controlled environment of test tubes (or their equivalents). The test tube contains only the components put there by the experimenters.

Experiments at the cellular level offer greater challenges to ensuring controlled conditions: an experiment performed with one cell type, for example, immediately raises the question of whether the results are valid for other cell types. Experiments with animals pose more questions of individual variation, and observations with humans often fall subject to problems with individuality. (This is why the extremes of psychology or social sciences have difficulty in being accepted as true sciences. It is hard to do a controlled experiment when every data point has a different basis.)

The use of *controls* is a distinctive feature of experimental science, and the reason why physics, chemistry, and (molecular) biology are regarded as "hard sciences." The first thing you look at in assessing a paper in experimental science is whether it has adequate controls. In effect, this means comparing the experiment, in which the parameter of interest is changing, with a control in which that parameter is fixed. If this condition can't be achieved, the experiment is not publishable. Practitioners of hard sciences harbor deep skepticism as to whether "soft sciences" (such as psychology or social sciences) should really be called sciences at all, largely because of the difficulty in establishing proper controls.

Biology has been an experimental science, practiced on a smaller scale than physics, in which experiments often need to be designed on a large scale to test theories. The practice of physics is divided between theoretical physicists and experimental physicists, but there is really no equivalent in biology, in which data rule the day and theory is not held in high esteem.

Physics has been occupied for decades by a search for a "theory of everything," and even though this might not be attainable, Nobel Prize–winning physicist Steve Weinberg could write a book entitled *Dreams of a Final Theory*.[9] Biology is more pragmatic. Since the discovery of the structure of the double helix and the breaking of the genetic code, advances in biology have been associated with ever-increasing amounts of data. Going into the twenty-first century, "big science" has invaded biology and brought its practice closer to that of physics.

Science has always been a rather self-contained system. Scientists accept, without thinking about it too much, that their mindset is different from the mindset of nonscientists. This means that some of the long-standing assumptions of science have rarely been questioned: that quality control is assured by the peer review system of asking other scientists to approve work before it is published; that including details of how experiments were performed makes it possible, in principle, to reproduce published work; that work is accurately reported and fraud or misrepresentation is rare; that science is intrinsically self-correcting because later work will show up errors in earlier work.

But the true scientific attitude should extend to questioning these internal assumptions. So I want to look at the validity of our long-standing assumptions about how science works, as well as to ask whether they remain true as we make a transition from science performed by small groups of individuals to science performed by large collaborative teams.

The pace of scientific and technological discovery has advanced rapidly over the past century in all fields of science, by any measure—number of scientists, number of published papers, number of new insights, and number of diseases that can be cured. Moore's "law," that the number of transistors in an integrated circuit doubles every decade, has held up well since it was proposed in 1965. Ray Kurzweil extends it in his argument that the Singularity is near (the point when computers can match human intelligence) to suggest that the rate of technological innovation now doubles every decade.[10]

Biology (at least in the sense of this book) is a more recent science than physics or chemistry and cannot claim to match that pace of technological advance. But in the time span covered by this book, the gaps between major discoveries have shortened, from four decades in the first half of the twentieth century to perhaps one decade at the end of the century. And the nature of discovery has shifted from the abstruse (what is the chemical basis of the genetic material) to the practical (how can we distinguish people by their DNA sequences). Molecular biology has made the transition from basic discoveries about living organisms to influencing daily life.

Changes in the concept of the gene are revealing about the nature of science as well as about the nature of the gene. The view of the gene as the sole unit of heredity dominated science until the concept of epigenetics introduced the idea that there might be other factors. Both epigenetics and the possibilities for gene editing created by the CRISPR technique have raised the question as to how far we are prisoners of our genes.

Should science be controlled—and, if so, how? Should scientists be completely free to tackle any problem, or are some techniques or experiments too dangerous or ethically questionable, so they should be subject to moratorium or permanent bans? The development of the CRISPR technique, rewarded by the most prestigious of all scientific prizes, the Nobel Prize, in 2020, raises a host of ethical questions going beyond science itself. But can there be an informed public debate without understanding of how science functions?

Increase in computing power has had a huge effect on all science. In increasing the pace of discovery, it raises the question of whether the traditional organization of science will be fit for purpose in the future. The recent extension from using mere algorithms into using artificial intelligence (AI) in research in biology calls into question whether the fundamental nature of research is about to change.

Although the issues I want to address are common to all areas of science, I approach them here through the prism of DNA. DNA is the thread of life both literally and metaphorically. If stretched out end to end, the DNA in one set of human chromosomes would be a very thin thread extending for about a meter; this is more than a million times the diameter of the cell that contains it. DNA is also the intellectual thread that holds together more than a century of scientific discovery, from Mendel to the latest results in molecular biology, from the concept that the gene is DNA to understanding inheritance, cancer, and evolution. In short, DNA is a unifying force in modern biology. The half-century following the discovery of the structure of DNA in 1953 saw an extraordinary flow of discoveries and surprises: can this continue?

This brings us back to the main question: what is science? At one point, I thought of calling this book *What Is Science?*, partly in homage to the book written by physicist Erwin Schrödinger in 1944, *What Is Life?*, which was so influential in persuading physicists and others to enter biology, inspiring the search to discover the physical basis for the genetic material.[11] That search led to the discovery of DNA: is it an exaggeration to say that our level of understanding of DNA is now so extensive as to become a dominant influence on human life? And DNA has become almost a catchphrase, with "it's in its DNA," meaning we have found the heart or quintessence of the matter. So this book is about the DNA of science: what science is, how it is practiced, and how that practice has been changing over the years.

Science depends on data, not on beliefs. I do not mean "belief" in any religious sense, but simply in the sense that even scientists may stick to ideas or hypotheses that are not actually in accordance with the data. One theme of this book is to show how ultimately data will triumph over mistaken beliefs.

I hold the view that science is intrinsically reductionist, but that leads us to ask what may be its limitations, not in the sense of questioning whether science can solve problems out of its sphere, but simply whether the reductionist approach can solve the ultimate scientific questions. This is as pressing a question now in biology as it has been in physics for the past half-century.

Almost a century later, I am not sure we are any closer to answering Schrödinger's question, *What Is Life?*, than when he posed it. Of course, Schrödinger meant it in a very precise sense: what is the nature of the hereditary material? With the immediate question of the genetic material resolved, biology can turn to beginnings and ends. So today we have progressed to asking the question in broader terms: how did life originate; and can we explain all functions of the organism, including human consciousness? It is a fair question whether these broader questions are answerable.

DATA

1

SCIENCE IN FLUX

Ask a scientist what he conceives the scientific method to be and he will adopt an expression that is at once solemn and shifty-eyed: solemn, because he feels he ought to declare an opinion; shifty-eyed, because he is wondering how to conceal the fact that he has no opinion to declare.[1]

Peter Medawar, 1969

S cience is poised at a point of transition. For several hundred years, there has been a continuing argument as to whether science proceeds from hypothesis to data (deductively) or from data to hypothesis (inductively).[2,3] But whether it works by either deduction or induction, or by both methods, there has been no dispute that data are valid only by comparison with controls (making it possible to identify the variables that change), and that correlations are not proof of cause and effect.

All this is swept away by "big data," which argues that obtaining a sufficient mass of data makes correlations sufficiently powerful in themselves. Big data by its very nature needs to be produced by big science, in which research is directed top-down in large teams of collaborators, rather than being conducted by small groups or individuals working independently. The two approaches represent different views of what is science.

In the modern era, ever since Karl Popper wrote his influential book, *The Logic of Scientific Discovery*, in 1934, it has been widely accepted that science proceeds by testing hypotheses.[4] More than a fashion, this dogma has dominated the view of science, with all alternatives indignantly refuted. The sheer power of the revolution brought by modern technology poses a revisionist question: is the standard of testing hypotheses still valid?

It may be a bit simplistic, if not idealistic, to suppose that scientists actually proceed exclusively by testing hypotheses. (Even more so to follow the strict Popperian view that the proper procedure is to try to falsify a hypothesis.) There is a good deal of action in the laboratory along the lines of, "that's a curious observation that doesn't fit our view of things, let's have a look and see what it means... ." It would be pushing things to argue that this rises to the level of trying to falsify a hypothesis.

Doesn't science really advance mostly by looking for data to amplify previous observations? And when there's a discord between theory and observations, you need data in order to formulate hypotheses. "Question-driven" science might be a more appropriate description of the approach than "hypothesis-driven" science, admitting that the "question" may vary from a vague curiosity to a precise hypothesis. Of course, no matter how research starts, by the time a paper is written, there will probably be a logical framework in place complete with hypotheses, as I show in Chapter 5.

Whether science proceeds by formally testing hypotheses, looking at curious discrepancies, or rooting about for data, at least until recently most scientists would accept a somewhat stereotyped view, emphasizing the importance of individual contributions. It's built-in to the way you become a scientist. Everyone starts as a PhD candidate when in principle you undertake a project that is sufficiently independent to be written up in a dissertation where you are sole author. Of course, a good deal of help and mentoring is required along the way. Then you obtain a postdoctoral fellowship. As a "postdoc," you join another laboratory, working under the guidance of the head of the laboratory.

All of this places a high priority on the ability to function independently and assumes that your contributions can be distinguished from those of others in the laboratory. This justifies the last step, which is to set up your own laboratory, where you will provide the basic intellectual direction and will recruit more junior researchers as postdocs and PhD candidates to develop your ideas.

Scientists have been very much committed to this (admittedly idealized) model of science, which really focuses on the role of the individual in conceiving projects to investigate. There was an uproar when it seemed at one point that funding for the Human Genome Project might threaten traditional funding of individual research in biology. It's a fair question whether a reflex like this represents rational defense of a great system for supporting science or Luddite resistance to the transition to a new system.

A successful project leads ultimately to a research paper authored by those members of the laboratory who have been involved. The paper is assessed by scrutinizing the data it presents, and asking to what extent they extend or change current thinking. This is different from the approach of big data, in which the "bigness" of the data means that the work is assessed by different criteria (and the real importance of the data may lie in its inclusion in a database).

The difference between question-driven science and data-driven science is typified by the attitude toward controls. Designing an experiment to answer a question requires the use of appropriate controls. An experiment is performed under two sets of conditions, which should be exactly the same except for the variable that is being measured. The experimental set has differences in the variable: the control set does not. For example, if we measured the effect of light on a phototropic organism, the control set could be dark.

There is really no equivalent in data-driven science. Big data takes the view that a sufficient mass of data compensates for any ignorance about exact sources, and controls (comparable to those used in question-driven science) become irrelevant. Data mining is a different scene from experimental science.

Scientists are always taught to be suspicious of correlations: big data turns this suspicion on its head. The case was put by a controversial article by Chris Anderson, the Editor of *Wired*, in an article in 2008 under the title, "The end of theory: the data deluge makes the scientific method obsolete."

The argument goes that "Petabytes allow us to say: 'Correlation is enough.' We can stop looking for models. We can analyze the data without hypotheses about what it might show. We can throw the numbers into the biggest computing clusters the world has ever seen and let statistical algorithms find patterns where science cannot."[5] Not surprisingly, this view was roundly trashed in the scientific community.[6]

As Mayer-Schönberger and Cukier say in their book, *Big Data*, "Big data is about *what*, not *why*. We don't always need to know the cause of a phenomenon; rather, we can let the data speak for itself."[7] You can't even call this approach data-driven theory; it is more like data-driven pragmatism.

Big Data points out that this replaces the scientific approach of looking for a representative set of sample data with simply looking at *all* the data. There's a shift in mindset from causation to correlation. A collateral issue is that the need for precision is overwhelmed by the quantity of data. "It is almost like a fishing expedition; it is unclear at the outset not only whether one *will* catch anything but *what* one may catch... . With so much data around ... hypotheses are no longer crucial for correlational analysis."

There is nothing new under the sun, of course. The debate between hypothesis-driven and data-driven science recapitulates changing views of scientific method. In inductive reasoning (historically) or in data-driven science (today), hypotheses are regarded as (if anything) damaging, because they may constrain the reasoning that should provide "hypothesis-neutral" conclusions driven directly by the data. Now we are back in the same debate.

Physics made the move from small-scale science to big science long before biology. In fact, we might ask if the laws of physics are an example of using big data on a nanoscale. Quantum mechanics implies a degree of uncertainty in descriptions at the subatomic level. But the laws of physics rule at a higher scale. Consider Brownian motion, which describes the random fluctuations in the movements of individual molecules in a gas. This makes the behavior of any particular molecule unpredictable, but the properties of the gas volume as a whole are entirely predictable. Uncertainty in the single datum is compensated by certainty when the mass of data becomes sufficient.

Even when science is performed in large groups or uses big data, the process is driven by human design. But the latest approaches supplement, or might even replace, human thought with artificial intelligence (AI).

Intuition is regarded as the key role in mathematical discovery, but is not sufficient to prove the most difficult theorems. Computerized analysis makes it possible to test relationships in large data sets, but in itself that essentially moves to a larger scale what used to be done manually. Using AI (more formally, machine-learning techniques) extends the use of the computer from increasing the data set to actually contributing to the hypothesis. However, in mathematics, to my mind, the results to date are more akin to using big data to spot undetected correlations rather than really to fall into the realm of AI (where the algorithm modifies itself).[8]

The more a field of science depends on algorithms, the less it's possible to understand it intuitively. Structural biology and DNA sequence analysis are the two areas in biology that depend most on computerization. I don't suppose crystallography is any more specialized than any other area of science, but it seems especially arcane to outsiders. "Only crystallographers understand crystallography. Everyone else has to take it on trust," one crystallographer admitted at a meeting I attended. But even crystallographers cannot take their software on trust.

Halogen azides are highly reactive explosive compounds. The structure of IN_3 (iodine azide) was analyzed by X-ray diffraction in 1993, and another form was discovered in 2012. The structures were resolved by computerized programs. But it turns out there are problems with the structures, including nitrogen atoms colliding with one another. The software missed some reflections that are apparent to the expert human eye in the original X-ray data. "The lack of crystallographic expertise of computers is ... a source of pitfalls," says Ulrich Müller, who worked on the analysis.[9]

Crystallographers are especially interested in finding alternative ways to analyze protein structure. CASP (Critical Assessment of Structure Prediction) is a contest, held every two years, in which participants submit models for proteins whose structures have been solved but not yet published. In the most recent round, CASP14, nearly 100 groups submitted more than 67,000 models for 90 target proteins. The software program AlphaFold was a standout.

DeepMind is a project to apply AI to science and medicine. It started in London and was acquired by Google in 2014. It developed the program AlphaGo that beat top players in Go. A game with defined rules might be regarded as close-ended, but protein folding is open-ended, so it was a significant extension when the program AlphaFold was designed to predict 3D shapes from protein sequences.

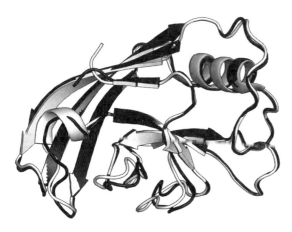

AlphaFold predicted a protein structure (black) that was in 93% agreement with the crystallographic structure (gray).

The first releases of its analysis in 2021 produced an impressive amount of data: structures of 20,000 human proteins, more than 300,000 proteins from all the key model organisms, and another 400,000 proteins from various other sources.[10] A year later, the next release extended the analysis to virtually all known protein sequences. The software was trained on structures that had been resolved, and validated by predicting the structures of proteins that had been solved but not yet published. Demis Hassabis, who cofounded DeepMind—"I'm half neuroscientist and half computer scientist," he says—points out that AlphaFold was based on structures that had been solved either crystallographically or by nuclear magnetic resonance (NMR)—"we're agnostic about methods," he says. "It learns about context, it infers it, even if it's not directly there, but what we'd really like to do is to model the full dynamics of protein [folding], including misfolding."[11]

The question posed by a true AI program (as opposed to very sophisticated algorithms) is whether it's possible to recapitulate the analytical process. Even its creators may not fully understand how it works. "To understand how AlphaFold predicts protein structure," they asked the software to identify the intermediate stages in analysis. This showed it works by an iterative process, continually improving the structure until it cannot go any further. "We hypothesize that [certain] information is needed to coarsely find the correct structure within the early stages of the network, but refinement of that prediction into a high-accuracy model does not depend crucially on [that] information" is the closest the authors come to understanding the software.[12]

The problem posed for the traditional view of science by a true AI program is that if we don't understand how it works, can we validate it theoretically? But does that matter if we can validate it operationally, by showing that it produces results as good as, or better than, conventional analysis? This may mean that crystallography becomes necessary only for confirming AlphaFold's results, and possibly at some point, confidence might be great enough to rely only on AlphaFold. "The [AlphaFold] system itself produces a confidence estimate, kind of a per residue confidence level—we thought that was important and it actually took us quite a long time to do—and [compared with experimental data] it's been spectacularly accurate. We knew if we wanted biologists to rely on it, they would need an easy way to [assess it]," Demis says.

The code for AlphaFold has been published as open source, so anyone can work on it, but the way it analyzes protein structure remains something of a black box. So how would we resolve a situation in which AlphaFold's prediction clashed with crystallographic analysis? Demis says that "Verification would be downstream, by using the program to make predictions, such as ligand binding, that ought to be borne out by subsequent data. [If the downstream effect doesn't work], then you would have to decide if the structure had been wrongly predicted or you'd done something wrong downstream. The dream would be that eventually you could rely on the computational analysis." Will we get to understand better how AlphaFold works with time? "It's just a temporary situation where we have pretty cool systems, but we don't understand them that well."[13]

By defining all the expressed genes in a particular cell type, and applying AlphaFold to the predicted protein sequences, it would be possible in principle to define the proteome (the entire protein complement of the cell) in terms of structures. Although this does not quite approach the Singularity (the point at which computers acquire consciousness and can compete with humanity), it certainly goes a long way toward replacing human researchers with automated protocols. It's a completely different way of doing science—in effect, replacing experimental research with theoretical analysis.

"There's a big question about AI in black boxes as to whether you can reverse engineer what they know. And of course, in the business of science, we're interested in how something works as well as the output it produces... . I would like to think [AlphaFold] is the proof of concept that computational analysis really can model biological phenomena. My dream would be to create a virtual cell, where you could perturb the cell and the model would give predictions of what would happen, without having to do painstaking years of experiments... . This is not just using AI to analyze data—It's different from that—this is AI modeling biological phenomena, actually building models of biological behavior. It's a new tool in the biologists' tool kit."

One of the major features of human-driven science is what you might call the intuitive approximation—simplifying data that are not exact to fit a more exact model. There is continual debate as to whether and when this is a necessary leap of inspiration and when it crosses a line into fraud. (I discuss this further in Chapter 10.) It's an irony that some AI programs actually get better when an analogous process is introduced to help them to deduce laws of physics from noisy data.[14] Is it sometimes necessary to introduce fudge factors to get closer to the truth?[15]

It is all very well to engage in a theoretical debate as to whether science proceeds by inductive or deductive reasoning, but just as "no plan survives contact with the enemy," none of this philosophy necessarily represents how science actually proceeds in the laboratory. The fact is that science is messy. Observations can be made because a researcher was testing a hypothesis derived from existing data, they can be made by accident or by mistake, or they can be made by trawling through a vast data set. There is no unique way of making discoveries.

Francis Bacon's view that science advances by experiments established the concept of the "scientific method" in the seventeenth century.[16] Calling a person who conducts experiments a "scientist" is more recent. In 1833, the poet Samuel Taylor Coleridge issued a challenge to find a name other than "philosopher" for people who did experiments.[17] The impetus was the prejudice of "real" philosophers against people who got their hands grubby with experiments, as opposed to *thoughts*. William Whewell (Master of Trinity College, Cambridge) defined the term in 1840: "We need very much a name to describe a cultivator of science in general. I should incline to call him a Scientist."[18] This reflected the movement from theorizing to experimenting, but many practitioners of science preferred anyway to be known as "natural philosophers," so the term "scientist" did not come into general use until toward the end of the century.

There is really no such thing as the much-vaunted "scientific method." It's a concept of philosophers of science, who debate what it might be, although working scientists would be more inclined to the admittedly circular view that "science, it turns out, is whatever scientists do."[19] In fact, working scientists regard the thoughts of philosophers about the scientific method with indifference if not disdain.[20] In a typical epithet, Richard Feynman (awarded the Nobel Prize for Physics in 1965) said, "philosophy of science is as useful to scientists as ornithology is to birds."[21] The (mistaken) view that there is a sole scientific method goes hand in hand with the mythic view of the scientific paper that I dismiss in Chapter 5.

Even if there is no distinct scientific method, science has a distinctive culture in which the validity of experiments is assessed, they are subject to reproducibility, and from there we work to establish chains of causality and predictability.[22] The distinguishing feature of a hypothesis in science (as compared with other fields) is that it can in principle be proved false.[23] Insofar as there is anything approaching a scientific method, it is that theory is subject to feedback, that is, modification by whatever results are actually obtained. Although there is no formal code of practice for scientists, equivalent to lawyers or doctors, following scientific methodology is the closest equivalent. Karl Popper called this the "scientific attitude."

The value system of science may be debatable, but the actual way scientists work day-to-day in the laboratory—what passes for scientific method, if it exists—is more controversial. Sociologists have looked at this also, but working scientists have about as much regard for the view of sociologists about science as they do for the views of philosophers: which is to say none at all. When anthropologists have studied science as a cultural entity from the outside, they have rarely understood it.

Attempting to study science from inside, Bruno Latour embedded himself in a laboratory at the Salk Institute for two years, like "an intrepid explorer ... living with tribesmen."[24] He produced a detailed account of day-to-day activities, but the anthropologist's bird's eye never synthesizes them into any real understanding of how science is conducted. He could not see the sum of the parts, let alone that the whole might be greater. It seems a fair question whether anthropologists' understanding of other cultures is any better, or whether they are limited by their own cultural methods and biases.

One of the difficulties in this approach was that it focused on "sequences, networks, and techniques ... [rather than]   individuals." Cataloguing daily activities rather misses the point. The most interesting aspect of the sociology of science is surely how individual behavior diverges from the idealized view of scientific practice. Science can be quite tribal, especially when it feels threatened—for example, by accusations of fraud. Reactions can owe more to loyalty to the culture than to logical analysis of data. It might be interesting to view science as a tribal society, or in some cases as a set of warring tribes, where each tribe runs like a fiefdom.

Scientists' skepticism about sociology is increased by the sheer nonsense of the "constructivist" approach. It would be hilarious, if it were not pernicious, to argue (as constructivists do) that science is a social construct in which the "facts" its practitioners observe are no more than artifacts of the methods used to analyze them. Postmodernism goes farther into arguing that all "facts" are relative to the observer. I mention this here simply because I do not want to be criticized for ignoring the view that there is another perspective from which to view science.[25] This idiocy achieved some prominence around the end of the twentieth century in what were called the "science wars."[26]

Put as kindly as possible, constructivism is an error showing that if you start with assumptions having no basis in reality, you end up with silly conclusions.[27] In the language of computers, it's GIGO (garbage in, garbage out). Ironically, nothing distinguishes the value system of science from other types of intellectual activity better than the self-referential nature of the constructivist or postmodernist attack on it.

When I came into science, the big distinction was between "pure research" and "applied research." Scientists engaged in pure research had the same disdain for those in applied research that natural philosophers had had for experimentalists a century earlier. The distinction had been drawn clearly by (once again) Francis Bacon, who complained that people "sought after *Experiments of Use* and not *Experiments of Light and Discovery.*"

In his lecture, *The Two Cultures*,[28] defining the split between science and the humanities, C.P. Snow lamented that, "Pure scientists have by and large been dim-witted about engineers and applied science... . They wouldn't recognize that many of the problems were as intellectually exacting as pure problems, and that many of the solutions were as satisfying and beautiful. Their instinct ... was to take it for granted that applied science was an occupation for second-rate minds."[29]

Today we would use the terms "basic research" and "applied research," and the distinction between them can be blurred, not least by the recent tendency of scientists to exploit their own discoveries (Chapters 3 and 4). I am afraid that with the increasing tendency to jargon, these are now often called "discovery research" and "translational research."

Irrespective of whether the rationale behind an experiment is testing hypotheses or trawling for data, there is a practical difference between a small group working on a problem, when each of them understands the whole gestalt, and a large group where the work has been broken into specialized tasks that are really understood only by those involved.

I thought it was a sign of the impending end of science when I had a discussion with an author about a gene-sequencing paper submitted to *Cell*, I think it must have been in the mid-1980s. Like a sentence in any language, a gene is read in only one direction. We call the start of the gene the "5′ end" (pronounced 5-prime) and the end of the gene the "3′ end." I asked one of the authors a question about the 5′ end. "Oh," she said, "I don't know about that, I'm the subcloner for the 3′ end." I thought that we had lost the plot if science had reached a point of specialization where you could know about one end of a gene and not the other end in your own research paper.

But there was worse to come. The review process at *Cell* was generally followed by a discussion with the author who had submitted the paper, often on the phone, and up to this time, they had usually been able to immediately answer whatever questions I or the reviewers had raised about the paper. Going forward, however, it became more common for the submitting author to need to consult with their coauthors before responding, and sometimes it became painfully clear that they did not in fact understand some aspects of their own paper.

The usual way scientists extend their research beyond their own expertise is to find a collaborator, but in the intramural program at NIH (the National Institutes of Health), contracts have been an alternative. When I was at the National Cancer Institute in the early 1970s, it was expanding rapidly as the result of Nixon's "War on Cancer," and some of the larger laboratories were able to expand their activities far beyond in-house research by using contracts. In fact, the atmosphere was a little like the Wild West, with contracts flying everywhere.

Using contracts doesn't seem unreasonable for developing some ancillary feature, perhaps refining an assay, but it's surprising for it to be used to obtain key data. The way contracts work was described by Matthew Gonda, a contractor who had taken the electron micrographs that were in two papers from the Gallo laboratory, published in 1984.

"It's a secret. It's their business. I get coded samples, I give them an answer. I don't have to know what it is. I give a viral diagnosis and that's all... . I didn't know what the samples were."[30] You might argue that this is a strength: effectively the micrographs are being analyzed double-blind, so there should be no prejudice. On the other hand, as a contractor, Gonda was not an author on the papers—which meant that none of the authors could vouch for a key point in the papers.[31] This is 180° away from the historic view that all the authors should be able to answer for all the data.

One effect resulting from the change to "big science" is that different skill sets may be needed to obtain the data and to analyze it. A whole new field of database analysis requires expertise in mathematics, modeling, and software that is quite different from the expertise required to obtain the data. This is a change in the practice of science, although it does not really address the principles by which it proceeds. The real issue is more the impact of increasing specialization on ability to think about wider issues.

Authorship of the paper reporting the sequence of the genome of the small plant *Arabidopsis* marked a change from the principle that authors take responsibility for their paper. Principal authorship was attributed to six "genome sequencing groups," with 109 authors between them. Another 39 authors were ascribed to one of 14 specific roles, such as comparisons with other genomes, analysis of specific parts of the chromosome, cellular organization, metabolism, response to pathogens, and so on. How can the team leader(s) have the expertise necessary to supervise all aspects of such a project?

The notion that each author on a paper has a full knowledge of its contents is beyond dead. Indeed, Drummond Rennie, the Editor of *JAMA* (*Journal of the American Medical Association*) was a strong advocate that the traditional list of authors should be replaced by an explanation of the responsibilities of individual authors. "The system of authorship, while appropriate for articles with only 1 author, has become inappropriate as the average number of authors of an article has increased... . Credit and accountability cannot be assessed unless the contributions of those named as authors are disclosed to readers, so the system is flawed," he argued.[32] He proposed that "authors" should be replaced by "contributors."

The argument has a certain force for papers in medicine, where there may be clear lines, for example, between researchers who obtain clinical results and those who perform statistical analysis on them. In fact, there is now a formalized system, called CRediT (Contributor Roles Taxonomy) for describing 14 specific roles that can be attributed to authors.[33] This is useful for research involving the large-scale collaborations of "big science." I am not so sure it applies so well to situations such as analyzing the two ends of the same gene.

Do large groups advance science differently compared with small groups? One idea is that "disruptive" papers, which introduce a new turn, will cite an earlier paper without citing many of its references. Papers making "incremental" advances will cite an earlier paper plus many of the papers that it cites. From this twist on citation analysis, which I discuss in more detail in Chapter 7, it seems that papers with five or fewer authors are far more disruptive than papers with ten or more authors. Papers associated with the path to a Nobel Prize are the most disruptive of all.[34]

Because large-scale science operates by consensus and involves the commitment of large resources, it has to follow predictable paths. Small groups are more likely to take risks, to follow hunches, and to make unexpected discoveries. Innovation depends on small-scale science, although there are some obvious exceptions, such as the need for large scale in physics to search for new particles.

"Research is being done quite differently now from what it was like some 75 years ago," says Nobel Laureate David Baltimore. "I am glad that I had the experience of a less controlled time in research... . When I started in science, people ran their labs in a very personal style. And even the trainees had their own way of keeping records and of doing experiments. Famously, some investigators kept notes on paper towels and many had most of their info in their heads, not on paper. There were famous investigators who sensed the way experiments were going rather than keeping any rigorous records. None of this is defensible but it was the reality. It meant that half-justified ideas got published and if someone found contradictory results, that led to progress. No one spent time trying to find out why earlier claims might have been faulty."[35]

If you were transported by a time warp back from a laboratory in 2023 to one in 1953, you would find yourself in a very different environment. In terms of the physical environment, almost every step in research would require manual intervention, compared with today's computerized equipment, omnipresent to the point at which you sometimes wonder how far human intervention is required. In terms of the social environment, the emphasis would be much more on individual efforts compared with collaborative work. I exaggerate a bit, of course, but you get the point that the environment has changed almost like comparing a workshop before the Industrial Revolution with a factory afterward.

Along with this change, there is a demand for more and more data to be included in a research article. "Science has changed in the way that suppresses publication of weird and wonderful findings," Nobel Prize winner Tom Cech says. "These days people don't judge a paper (just) on its own merits, but want to go further. Sometimes it's useful, but sometimes it's suppression, and journals have taken on the role of ethics police. Some of it is good, but sometimes it becomes so controlling."[36]

Science is technically better today than ever before. Although its value system remains the same—based on verification and self-correction—it's a fair question what effect the change from individuals to groups is having on the way researchers practice science. Ironically, one effect of the move toward big science is that errors produced by excess devotion to dogma, which I discuss in Chapter 15, become irrelevant when the focus is on correlations instead of theory.

Data, no matter how unlikely or inelegant, will always trump theory. This is part of the ying and yang between hypothesis-driven and data-driven science. A fair way to put the issue might be that all science is data-driven and the question is to what extent the search for new data should be driven by hypothesis.

The objective of science remains the same—producing data for others to scrutinize—which speaks to something fundamental about the scientific attitude. The process of discovery has changed in some cases from looking for cause and effect to seeking correlations, and the process of presenting results has moved toward databases. But as the French would say: *plus ça change, plus c'est la même chose.*

But it is unclear whether we have yet seen the full effects of the move away from hypothesis-driven science toward data-driven science, and whether we are about to enter a new era. And if the past decades have been marked by a move from hypothesis-driven science to data-driven science, will the future see a move from human-driven science to AI-driven science?

CHAPTER
2

DATA MINING

It is a myth that the success of science in our time is mainly due to the huge amounts of money that have been spent on big machines. What really makes science grow is new ideas, including false ideas.[1]

Karl Popper, 1975

Obtaining data has always been part of science. John Sulston, who later became the director of the British contribution to sequencing the human genome, started his career by mapping cell lineages in the worm. "I shut myself away and devoted myself to the embryonic lineage... It would take a year and a half of looking down the microscope every day, twice a day, for four hours at a time... My first studies of the worm lineage didn't require me to ask a question ... they were pure observation, gathering data for the sake of seeing the whole picture."[2]

What could be superficially more different than one person sitting at a microscope, and a vast team of technicians running DNA sequencing machines? Yet the objective is strangely the same: to gain a basic set of data. But there is a great difference between seeking further data to help investigate a specific problem, even if a hypothesis is not formally being tested, and simply trawling for data in the hope that something interesting will emerge. The latter type of approach would have been dismissed as a fishing expedition, and never approved for funding historically, but it may be becoming the rule rather than the exception today.

What I'm concerned with here is not so much the difference in *modus operandi* between question-driven and data-driven science as such, but what effect this has on the intellectual atmosphere, on the ability to conceive projects that will take science forward to the next stage of understanding.

The Human Genome Project, discussed in Chapter 17, was in a sense the biggest fishing expedition in the history of biology. It did not set out to test a hypothesis; it was an overt attempt to collect a massive amount of data that would be useful for future studies. Although many forceful arguments for the intellectual interest of the project were advanced at the time, you have to wonder whether part of the rationale was the same as George Mallory's answer when he was asked why he wanted to climb Mount Everest: "because it's there!"

A genome sequence is a fantastic tool, but it is a means, not an end. Progress to the next stage of asking interesting questions about biology or solving medical problems requires a transition back to hypothesis-driven science. Suppose sequence data suggest a correlation between a particular genetic makeup and cancer. To cure the disease, we need to formulate a hypothesis about the action of the gene(s) involved, and test ways of modifying their actions. Of course, in this brave new world, there might now (in principle) be the alternative of simply changing the genome to a combination of sequences known not to be involved in cancer, without asking any further questions... This would be an example of the "big data" approach.

Google's replacement of epidemiology by using search data to analyze the spread of influenza has become a classic example of the application of big data to biology.[3] Looking back at online queries between 2003 and 2008, a set of 45 search queries (out of 50 million search queries) correlated with the occurrence of influenza. In principle, this should enable future epidemics to be tracked in real time, providing better forecasting than could be done by traditional epidemiology. The striking feature here is that the algorithm does not depend on any analysis of symptoms related to flu, or even any knowledge of the characteristics of the disease, once the basic correlation has been established.

Google Flu Trends sounds brilliant: but from 2011 to 2013, it greatly overestimated the trend, and estimates from the U.S. Centers for Disease Control and Prevention (CDC) (based on traditional epidemiology) did far better.[4] This highlights a problem with big data: without understanding the algorithm, it's impossible to say whether aberrant results mean the algorithm is simply invalid, or perhaps just needs tweaking or better training. If there's no way to verify the results, how does the self-correcting mechanism of science apply? Perhaps science will have to change its criteria to accommodate big data.

We had a fairly brutal test at *Cell* for applicants for positions as scientific editors. We would give them a copy of the latest issue and say: walk us through this, explain the logic of each article, whether the data support the conclusions, and whether we were right or wrong to have published it. When I started doing this in the early 1970s, a fair proportion of applicants made a good stab at logical analysis and assessing the general interest of a paper. By the time I left at the end of the 1990s, fewer and fewer applicants could handle this successfully, although they would often make excellent detailed comments about any technical aspects of a paper that they were familiar with. I worried that scientists were turning into superb technicians at the expense of broader understanding. How far can techniques and data mining take us?

Once sequence data have been deposited in a data bank, they are effectively free for anyone to interpret. Indeed, free access is the intention of the Bermuda Principles established during the Human Genome Project, which require new sequences to be rapidly deposited in a database, as I discuss in Chapter 6. An early demonstration of the power of the database to establish connections came in 1983 when Russ Doolittle searched the protein database at the University of California, San Diego, and found a relationship between a protein of the simian sarcoma virus and the human growth factor PDGF.[5] He had not previously been involved with the work. Here was a connection between cancer and growth factors that had escaped the researchers working on the two proteins.

Today, anyone working on a specific protein or gene would routinely check the databases: there would be no opportunity for interlopers. Yet the existence of databases always leaves open the possibility to develop new means of analysis that may show new connections. Certainly this enlarges the possibilities for reinterpretation of data.

In a strange throwback to the attitude that researchers are entitled to a stranglehold on specific problems (discussed in Chapter 13), an editorial in the *New England Journal of Medicine* as recently as 2016 expressed grave concern about the use of open data. "A ... concern held by some is that a new class of research person will emerge—people who had nothing to do with the design and execution of the study but use another group's data for their own ends, possibly stealing from the research productivity planned by the data gatherers, or even use the data to try to disprove what the original investigators had posited. There is concern among some front-line researchers that the system will be taken over by what some researchers have characterized as 'research parasites.'"[6]

Surely this is an attitude more typical of Big Pharma than of science (and perhaps an insight into how Big Pharma has corrupted the practice of medicine?): the very thought that another researcher might try to disprove your theory—*quelle horreur*! More seriously, the whole point of science is that you contribute your data to the public domain, so that they can be tested by others. Roll on the "research parasites."

Obtaining "big data" requires big groups. A vivid sign of the way science has changed is the number of authors on a scientific paper. This has increased from an average of less than five in the 1990s to 10–20 today in biology. It is more in physics. The extremes are presented by papers a century apart. Ernest Rutherford reported the discovery of the proton in 1911 in a paper on which he was sole author. The two papers reporting the discovery of the Higgs Boson in 2012 each had a thousand authors!

If a paper has 10–20 authors, it is unlikely that many of them will really have a grip on the whole thing. It is as though science has changed from performance by individual soloists to performance by an orchestra in which only the conductor has the whole picture.

Biotech led the way, with its need to combine disciplines in order to reach the defined goal of producing a specific product. "Out of Genentech came papers with twelve or fifteen names on them, and it was always viewed by academe as a funny way of doing science. I found to the contrary it was a very efficient way of doing science, because this was a demonstration that you can accomplish a lot by working together with different disciplines," says Herbert Heryneker, who was one of the early researchers to work at Genentech.[8] Today this has become normal in research.

Beyond that, there can be active disapproval of scientists who work alone. Howard Green, a cell biologist whose long distinguished career started in the 1960s, maintained a small laboratory, first at MIT and then at Harvard Medical School, but right up to the end of his career was involved in the laboratory hands-on and even published the occasional

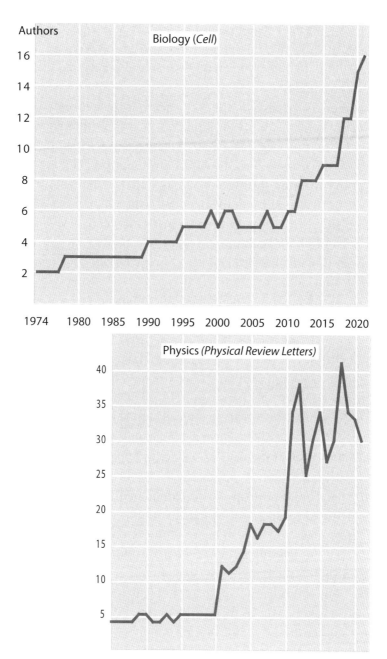

Papers in biology and physics both had up to five authors in the 1980s and 1990s and then the number increased. Biology papers now usually have 10–20 authors (as seen in Cell*), whereas physics is "bigger" science, with 30–40 authors per paper (as seen in* Physical Review Letters*).[7]*

paper as sole author. When I published one of these papers in *Cell* in the late 1990s, a Nobel Prize winner at MIT said to me disapprovingly, "he really shouldn't be doing his own research and publishing like that, he should be running the lab."

That there is a change in the way science is done is undeniable: the question is what it means. Is the argument about the difference between hypothesis-driven science and data-driven science more than semantics? "'Discovery-science' has absolutely revolutionized biology," says Lee Hood, at the Institute for Systems Biology that he founded in Seattle. "It's given us new tools for doing hypothesis-driven research."[9] Formerly at Caltech and then in Seattle, Lee was instrumental in developing automated sequencers for both protein and DNA; he was a major player in the Human Genome Project, and is a strong advocate for the use of big data in biology.

This approach is now sometimes called "omics," referring to use of databases of the genome (genomics), proteins (proteomics), metabolites (metabolomics), and so on. Some people even talk about the "omics revolution." It has converted some skeptics. David Botstein, formerly of MIT, then at Princeton, and originally a skeptic of the Human Genome Project, was coauthor of a paper on DNA microarrays (an early method for identifying many genes at once) that said, "This process is not driven by hypothesis and should be as model-independent as possible."[10] This is a complete reversal of the view of science as hypothesis-driven.

Several of the "omics" approaches have been combined into the ENCODE database (Encyclopedia of DNA Elements), founded by NHGRI (National Human Genome Research Institute)—if you want to follow science these days you have to know all the acronyms—which is a database focused on defining all the elements in the human genome, both coding for proteins and regulating gene function.[11]

Will multiomics be the approach to medicine in the next century? Lee Hood thinks so. He testified to Congress in favor of a Beyond the Human Genome Project to use all sorts of omics—not just genomics and proteomics but also a data set including basically anything that can be measured—to assess "wellness."[12] Looking at blood proteins, for example, Lee believes that some diseases can be detected years before any clinical manifestation. This resembles big data in seeking correlations in a vast data set, but here the data set is related directly to factors known to be involved in human health.

The sequence of the human genome is the most obvious example of big data in biology. Or perhaps that should be plural, human genome sequences, because the "bigness" of the data is partly due to the number of genomes that have been sequenced. Attempts to define human conditions that have been intractable to conventional analyses have tried using an adaptation of big data with the "omics" approach, in effect replacing any effort to define cause and effect with the power of mass correlations.

Could correlations among sequence patterns identify susceptibility to disease? It turns out from analyzing genomes of 3000 patients with COVID that there's a region on human chromosome 3 (containing six genes) that influences the severity of infection. One particular sequence variant is strongly associated with the risk of developing severe enough symptoms to require hospitalization, and turns out to have been inherited from Neanderthals![13] Here is a demonstration of the potential of big data to develop diagnostic insights that could not come from traditional science.

Ability to perform genome-wide methylation analysis has led to studies to try to correlate the methylation pattern with depression, schizophrenia, or obesity. (Methylation is a process associated with turning genes off. It's often considered to be an epigenetic mechanism, meaning that it identifies effects that happen at the level of the genome, but that are not due simply to the sequence of DNA. Chapter 19 shows how the fashion for epigenetics has led to experiments such as these.)

This is certainly not hypothesis-driven science, but nor is it quite big data either, as it's not fully independent of context: the basic assumption is that methylation patterns may identify genes that are differently expressed in a disease, which could be a starting point for investigation. Of course, you have to know which tissue or cell type to investigate. A typical study along these lines identifies tens of candidate genes.

The chance of identifying an individual gene of importance is low. The approach reminds me of the old story about the drunk looking under the streetlamp for a coin he had dropped. "Why are you looking under the lamp when you dropped the coin elsewhere?" "Because that's where the light is." It's possible to use big data thoughtfully or thoughtlessly.

3

PATENTS VERSUS SCIENCE

Edward Murrow (1955):   *Who owns the patent on the polio vaccine?*

Jonas Salk: *Well, the people, I would say. There is no patent. Could you patent the sun?*[1]

Everywhere in science the talk is of winners, patents, pressures, money, no money, the rat race, the lot; things that are so completely alien … that I no longer know whether I can be classified as a modern scientist or as an example of a beast on the way to extinction.[2]

June Goodfield, 1981

Does the value system of science conflict with the patent system? Science requires a free exchange of information, "free" meaning that the information is both available and without charge. And once research is published, it is part of the scientific canon: no one owns it. Patents require the information supporting the patent to be revealed, but give the patent holders exclusive rights to the discovery, allowing the inventors to recoup their investment by charging for use of the patent. There is no requirement that a patent holder actually licenses an invention: it can be used to maintain a monopoly.

As exemplified in the U.S. Constitution, the original idea behind the patent system was to "promote the Progress of Science and useful Arts." The question is whether patents now assist or impede scientific research?

Patents have always been part of physics or chemistry, but less so in biology. They can apply to methods or to objects. An invention must be novel (no one else has reported it previously), which gives rise to furious fights about priority. They must also be useful, and "nonobvious," which causes arguments about whether one discovery is a predictable consequence of another. It's a condition of granting a patent that it enables a person with ordinary skill in the field to make or use the invention without undue experimentation.

This is not so different from the principle of reproducibility that is supposed to apply to scientific papers: in fact, it is a pity that science does not have an enforcement mechanism equivalent to the patent's—denial of validity. (This could take the form of a formal mechanism requiring papers to be retracted if authors were unable to respond in a satisfactory way to attempts to reproduce them.)

There are few limits on what can be patented. One of the exclusions is "products of nature," but there is a giant loophole here that goes back to the start of the twentieth century, when a patent for adrenaline was supported because it was isolated and purified away from its natural surroundings. The fiction that a purified product is different from a natural product has reverberated ever since, all the way to the question of whether DNA sequences can be patented.[3]

Exclusion of products of nature was generally taken to apply to patenting of living organisms, and indeed, in 1973, the U.S. patent office rejected an application to patent a bacterial strain. Ananda Chakrabarty at General Electric applied for a patent on *Pseudomonas putida*, which he had engineered with genes coding for enzymes that attack hydrocarbons. After a series of lawsuits, almost a decade later, in 1980, the

RECOMBINANT DNA 101

The discovery of recombinant DNA spawned the biotech industry in the 1970s and was at the heart of several major decisions on patents. Discovery is not really the right word, because recombinant DNA is an artificial construct, created by joining together DNA sequences from different origins. The principle is to cut two molecules to generate ends that are complementary so they can cross-join. There are now several methods to do this. *Cloning DNA* is a shorthand description for using recombinant DNA to produce large amounts of a target sequence by linking it to the sequences that enable it to be easily reproduced by the process of DNA replication (most often in a bacterium). A *cloning vector* is an independently replicating molecule of DNA with a site where a target sequence can be inserted in order to be cloned.

Supreme Court issued its famous ruling upholding the patent. "Anything under the sun that is made by man," can be patented.[4]

This was surprising but not particularly threatening to research: *P. putida* was designed for a specialized commercial interest, and would not be likely to attract myriads of scientists who wanted to work on it. Controversy focused on the general question of whether living organ isms should be patented. With awareness of patents heightened by the decision, the patent wars moved on to the field of everyday techniques or information.

At the same time, Chakrabarty was engineering *Pseudomonas*, a collaboration between Herb Boyer at the University of California and Stan Cohen at Stanford University was cloning recombinant DNA. (Recombinant

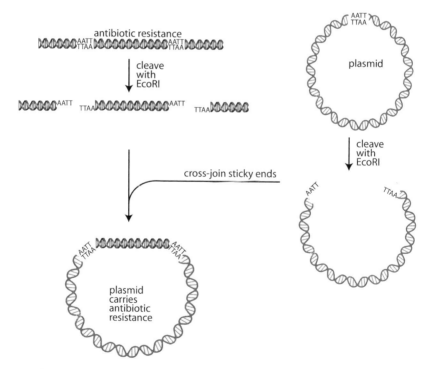

An early method for cloning DNA. A "restriction enzyme" called EcoRI (a specialized activity that makes breaks in DNA) cleaves DNA containing a gene for antibiotic resistance (left) and a circular plasmid (right) to create "sticky ends." Cross-joining the ends inserts the antibiotic-resistance gene into the plasmid. The plasmid replicates when it is introduced into a bacterium and can be retrieved by selecting bacteria for resistance to the antibiotic.

DNA became the accepted term used to describe a DNA that had been artificially created by joining together two other DNA sequences.) Boyer had developed methods for cutting and joining different DNAs together, while Cohen had developed a system in which DNA could be reproduced by incorporating it in a plasmid (an element that can replicate in a bacterium independently of the chromosome). Their first experiment incorporated a gene for antibiotic resistance into a plasmid, which then converted a bacterium to antibiotic resistance.[5] A year later, they introduced a gene from frogs into *Escherichia coli*—the standard bacterium for research—and now it seemed possible that the system could be used to produce proteins for medical use, such as insulin.[6]

This was immediately recognized as a powerful method for cloning (potentially) any DNA. The possibilities came to public attention when the *New York Times* published a report under the innocuous title "Gene transplants seen helping farmers and doctors."[7] Stanford University had an unusually active patent office, where Niels Reimers, who had founded Stanford's technology commercialization program in 1970, was stimulated by the report to urge the authors to apply for a patent. "My first reaction was quite negative," Cohen recollects. "Could findings of basic research funded by the public be patented, and should they be?"[8]

That question was resonating in the scientific community. Niels Reimers recalls dryly, "There was not universal joy in the biochemistry establishment about what I was doing."[9] Boyer and Cohen encountered some hostility when they talked at meetings. "We gave talks, and people were jumping up and yelling, 'Is it true you're patenting recombinant DNA? How can you do that?'," Herb Boyer recollects.[10]

In the end, there was a scramble to file the patent application before the deadline imposed by a requirement that applications (in the United States) must be filed within a year of any public report. (In other countries, patent applications must be filed before any public report, creating a dilemma as to whether to publish in order to get scientific priority or risk being scooped while delaying to make the patent application.)

Stanford filed for a group of related patents. (Herb Boyer's institution, the University of California, agreed to go along, but was less enthusiastic: it did not even want to pay its half of the filing fee.[11]) After various technical delays, the patents were issued in 1980. Reimers established an unusual strategy for licensing that became a model in biotechnology. The patent was licensed nonexclusively, meaning it was available to any company that was prepared to pay the fees and royalties (as opposed to granting an exclusive license to one company). And there were no fees for academic use. The patent generated $255 million for the universities over its 17-year lifetime.

Like most scientists at the time, Herb Boyer and Stan Cohen had little knowledge about patents and no interest in them. If it had not been for Niels Reimers overriding their (mistaken) view that the work could not be patented because it had been performed with grants from the National Institutes of Health (NIH), history might have been different.

Before 1980, it required special procedures for universities to apply for patents on work performed with federal funds. Reimers is a staunch defender of the Bayh–Dole Act of 1980, which allowed universities (and nonprofit organizations) to patent discoveries made with federal funding. In fact, subsequent changes in the law encouraged them to do so.

"When I joined Stanford University in 1970 ... most university advances backed by federal dollars never left the laboratory. Patents simply reverted to the government, which generally did little with them. When lawmakers drafted Bayh–Dole, fewer than 5% of federally held patents had been licensed to private companies. The new law unlocked those patents' potential... . This single reform created an unprecedented wave of private-sector innovation. Between 1996 and 2017, more than 13,000 start-ups formed based on licensed university research," Reimers says.[12]

The tide of opinion among scientists has swung to and fro over whether patents are good or bad for science. When César Milstein created monoclonal antibodies in 1975, failure to patent the method became a source of controversy. When the NIH filed for patents on gene sequences in 1991, there was widespread disquiet.

César Milstein was working at the MRC Laboratory of Molecular Biology in Cambridge (England) when he and Georges Köhler developed a technique for fusing antibody-producing cells with cells from a cultured myeloma to produce hybridomas—cells that produce a single antibody and grow indefinitely. Both Milstein and the MRC were aware of the commercial possibilities. Sending Tony Vickers, their contact at the MRC, a preprint of the paper that was in press in *Nature*, Milstein

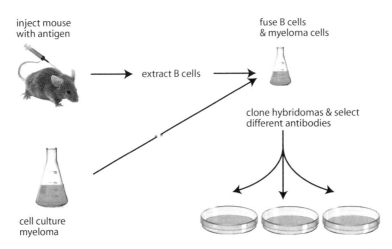

inject mouse
with antigen

extract B cells

fuse B cells
& myeloma cells

clone hybridomas & select
different antibodies

cell culture
myeloma

Making monoclonal antibodies. Cells making antibodies (B cells) are extracted from the mouse and fused with cells cultured from a myeloma (a cancer of the blood cells). Fused cells can be cloned to give cell lines (hybridomas) that produce specific (monoclonal) antibodies.

wrote, "The interest of such cell lines depends on the possibility of producing antibodies of potential practical value. The great advantage of cell cultures is that one can not only standardize their product but also obtain antibodies of the monoclonal type, i.e. of a uniquely defined specificity."[13] Köhler and Milstein shared a Nobel Prize in 1984.

The paper was not given great prominence when it was published and did not attract much interest in the press. Milstein recollects that, "It was submitted as an article, that is, longer and complex, but [*Nature*] replied yes, it was acceptable but not of real general interest, so they would not publish it whole. We cut it to 1500 words for a letter and it was a big problem cutting this down."[14] The only immediate press report was on the BBC World Service.[15]

The problem at the time was that the MRC did not have the ability to apply for patents but had to do so via the National Research Development Corporation (NRDC), which in typical bureaucratic fashion could not be persuaded to act. Time was of the essence because, according to British patent law, the application had to be submitted before the paper was published in *Nature*. Even a year later, in retrospect, the NRDC could not see the point of it. "It is certainly difficult for us to identify any immediate practical applications which could be pursued as a commercial venture ... and it is not immediately obvious what patentable features are at present disclosed in the *Nature* paper."[16] When the missed potential became a scandal at the end of the decade, the scientists were blamed for their "lack of appreciation of the NRDC"![17]

The scandal was sharpened by both looking back and looking forward. There was criticism that this was a repeat of the opportunity lost when the discovery of penicillin was not patented—déjà vu all over again. Even if the scientific community may have been ambivalent about patents, by this time there was no doubt about their significance at the MRC.[18]

The scandal was triggered in 1979 when a group at the Wistar Institute in Philadelphia obtained a patent on monoclonal antibodies. Milstein had sent the myeloma cell line to Hilary Koprowski at the Wistar in 1976, with the proviso that it should not be distributed further or used for commercial purposes. "Essentially they are patenting our procedure," Milstein said.[19] He added ruefully, "We were too green and inexperienced on the matter of patents... . We were mainly concerned with the scientific aspects and not giving particular thought to the commercial applications." The patent was granted in the United States, but not in Britain. Anyway, it was generally ignored.

Attitudes in the scientific community began to change as the result of the success of the patent on recombinant DNA, and the failure to patent monoclonal antibodies, and scientists came to accept that there were advantages to patenting innovative methods. Even in the European system, in which publication cannot precede the patent application, it was possible to work around the restriction and avoid delaying publication. David Secher, who filed an application in 1980 (against the advice of the MRC) to patent a hybridoma making interferon, recollects that "we paid only £5 to deposit the *Nature* manuscript only in the Patent Office. We delivered the manuscript to the *Nature* offices on the same afternoon. The lawyers laughed, but it has held good."[20]

The range of things that could be patented had expanded greatly by 1980. With the granting of the patent for recombinant DNA, and the decision on Chakrabarty, it was clear that methods, cells, and even organisms were eligible for patenting. But no one had even contemplated the possibility of patenting sequences of DNA.

It was a bombshell when Craig Venter mentioned in a congressional hearing in 1991 that the NIH was planning to file patent applications on the expressed sequence tags (ESTs) he was sequencing in the human brain at the National Institute of Neurological Diseases and Stroke (NINDS). ESTs are short DNA sequences, made artificially, that represent the set of genes that are expressed; they could be used as a shortcut to select genes for sequencing. The proposal to patent them was only the first of several provocations that would come from Venter during the Human Genome Project.

The universal reaction was that sequences probably could not be patented; and even if they could, they shouldn't be.[21] The strength of the reaction is indicated by the fact that protests came even from the advisory committees to government agencies funding the Human Genome Project. Nobel Prize winner Paul Berg, Chairman of one committee, said, "It makes a mockery of what most people feel is the way to do the genome project."[22]

The impetus came from Reid Adler, in the Technology Transfer Office at the NIH, who made the traditional argument for patents. "If everything goes into the public domain, there is much less incentive for companies to invest time and money in developing a project." It was certainly questionable whether DNA sequences with no known function could satisfy the criterion of usefulness for a patent, but Adler argued that it was time to test the question of patentability, and that if the NIH failed to do this, it might in effect lose control of its own work.[23] Bernadine Healy, Director of NIH, who had not been consulted about the application, defended it: "We decided to file in order to hold our place ... because no one knows to what degree function must be tied to structure for the patent requirements to be satisfied."[24]

The announcement ran the risk of starting a patent war. The British announced that they too would file to patent sequences (and because some were bound to overlap with Venter's sequences, this would imply direct conflict). The French got the European patent office to say that the sequences could not be patented in Europe.[26] This was a great contrast with the international cooperative spirit of the Human Genome Project.

The original patent application (on 315 sequences) in 1991 and a second application for another 2106 sequences in 1992 were rejected;

the NIH appealed, adding another 4449 sequences, and was given a year to appeal after this was rejected in 1993. After Harold Varmus became Director of NIH, in 1994 he withdrew the application; and the British followed suit immediately.

This was far from the end of proposals to patent sequences. Craig Venter left the NIH to found the nonprofit Institute for Genomic Sciences, where he thought he would be able to do sequencing at a much faster rate. Becoming a leader in the field, in 1998 he teamed up with Michael Hunkapiller of Applied Biosystems, the leading manufacturer of sequencing machines, which was owned by PerkinElmer. The new venture, called Celera, proposed to sequence the human genome within the next three years on a commercial basis. (I discuss the outcome in Chapter 17.)

The outcry was summarized by John Sulston. "The whole future of biology came under threat. For one company was bidding for monopoly control of access to the most fundamental information about humanity."[27] The notion that payment might be required to access the genome sequence ran counter to the whole value system of science. (Of course, paying for information is not unprecedented: most research articles were published in journals that are available only by paid subscription—but the scale was a bit different here.)

Celera made its money by selling access to its database, until revenues declined to the point at which this was no longer profitable.[28] The principle became established that sequences could be patented; even those with unknown function could be patented as "probes" for gene function.[29]

The floodgates opened, and to date more than 350,000 patents have been awarded for DNA or genes, including more than 250,000 that mention DNA sequences. The number of patents for processes based on the CRISPR gene-editing technique has increased from 1 in 2011 to 2000/ year today.

Who owns these patents? Pharmaceutical companies own almost half, and biotech companies own another quarter.[30] Universities and the NIH own the remaining quarter.[31]

Allowing DNA sequences to be patented created collateral damage for the researcher, who can be shut out of his own sequence if someone else later discovers a reason to patent it. John Sulston described the reaction

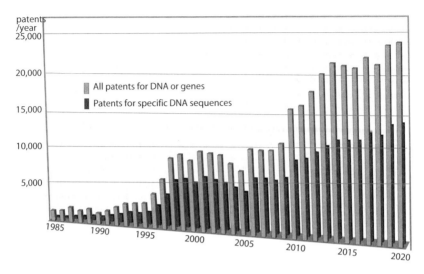

The number of patents for DNA or genes rose from just over 1000/year in the 1980s, passing 10,000 in 2006, and 20,000 in 2014. Only 100 mentioned DNA sequences in 1985, but today these are more than half of all DNA patents.[25]

at the Human Genome Project. "One of the aims of the HGP has been to 'raise the bar' by making as much genome information as possible universally available in the public domain and therefore unpatentable."[32]

That argument took an unlikely turn when major drug company Merck spent several million dollars funding Washington University in St. Louis to perform sequencing and deposit the results in a public database. The rationale was that this would make the sequences available to all, and would protect Merck from being forced to pay fees to the start-up companies (in particular, Human Genome Sciences and Incyte) that were engaged in genomic sequencing at the start of the 1990s.[33]

But even when sequences are saved in a public database, it's still possible to patent variants with particular uses. Sulston described the effect on research. "The Sanger Center began life in an environment in which commercial pressure was always going to be part of the picture. Those who were working to map particular human genes either expected to secure patents on them, or were terrified that someone else would beat them to it. It made for an atmosphere of mutual suspicion." (The Sanger Center was established under Sulston as Britain's major DNA sequencing center.)

Most patent fights involve conflicts either between overlapping patents held by different parties, or challenges when a patent is ignored. They are most often settled before the case goes to court, but some have involved protracted battles going all the way to the U.S. Supreme Court. The arguments seem to be more intense than those about scientific priority, perhaps because either you get the patent or you don't, whereas with primary research you can always continue to argue your case and to publish further data. And, of course, substantial sums of money can be involved with patents.

The patent situation changed dramatically in 2013 when the U.S. Supreme Court ruled on the challenge to a patent on the *BRCA1* gene. There was a lot at stake in this lawsuit, which effectively challenged the idea that a human gene could be patented. By 2005, patents on DNA sequences that mentioned specific human genes applied to about 4000 genes, roughly 20% of the total in the genome.[34]

When Mary-Claire King started working on breast cancer in the 1970s, there was no evidence for a genetic basis. By the 1980s, familial patterns of breast cancer had been established, leading to the deduction that this was caused by a gene, which she called *BRCA1*, on human chromosome 17. It could be responsible for 5%–10% of breast cancers, and is especially pronounced in Ashkenazi women. The race to identify the gene started in 1991.[35] It was identified by a group led by Mark Skolnick at the University of Utah in 1994.[36] *BRCA2*, which is responsible for a smaller proportion of breast cancers, was isolated in 1995.

Subsequent history shows the tangles that can be created when researchers have two hats, working at a university with public funds, such as NIH grants, but also involved in a private company that intends to exploit the discoveries made at the university. Although this type of relationship has always been encouraged by the entrepreneurial attitude at, say, MIT or Stanford, these were exceptions to the general policy at most universities, until becoming the norm in biology in what we might call the genomic era. I discuss the role of biotech in the next chapter.

Authors on the *BRCA1* cloning paper came from three sources: the University of Utah, NIEHS (part of the NIH), and Myriad Genetics, a company that Mark Skolnick had cofounded. The NIH assigned its patent rights to the University, which gave exclusive rights to Myriad, which obtained a slew of patents on both *BRCA1* and *BRCA2*. The controversy started when Myriad offered a diagnostic service at what was regarded as prohibitively high price (around $3000 per test)[37] and aggressively enforced its exclusivity by preventing anyone else from offering diagnostics, or sometimes indeed from even working on the genes. This was an extreme business model and brought the house crashing down.

Indignation that only those women whose insurers were prepared to pay the fees could use the diagnostic service led the American Civil Liberties Union (ACLU) to bring a lawsuit (against Myriad and the U.S. Patent Office). The basis of the action was a claim that a human gene could not be patented, under the guise of a complaint about damage suffered to patients and others because of the patent. (Presumably the ACLU was equally indignant about the predatory pricing practiced by drug companies in the United States generally, but did not feel there is a constitutional issue involved.)

The initial ruling on the lawsuit invalidated the Myriad patents, the appeals court reversed the ruling, and finally, in 2013, the Supreme Court ruled that human genes could not be patented. This invalidated more than 4000 patents that the Patent Office had already granted. The judgment drew a peculiar distinction between the DNA sequence of a gene (which could not be patented) and the sequence of cDNA (which could be patented). A cDNA is a copy of the sequence of the gene after it has been expressed (with nonfunctional parts removed). To a scientist, in terms of patentability, this is a distinction without a difference; but it fits the legal fiction that cDNA is man-made, whereas a gene's DNA occurs in Nature.

Courts are not well-equipped to understand molecular biology; struggle as they may, lawyers make comical misunderstandings about science. If the scientific community had been left to make the judgment, it is unlikely that DNA sequences representing genes in any form would have been regarded as patentable. But maybe scientists do not understand patent law any better than lawyers understand science.

Patents are not good for science, but they are not necessarily bad for it. There is no advantage for science as such in protecting a discovery with a patent; this is basically irrelevant to the process of intellectual discovery. The purpose of promoting technological development of a

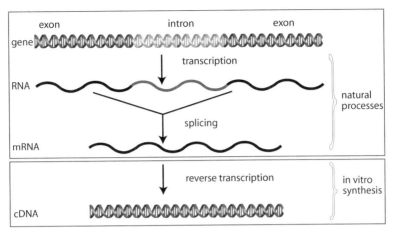

Did the Supreme Court misunderstand DNA? A gene is a continuous sequence of DNA, divided into alternating parts called exons (black) and introns (gray). When it is expressed by "transcription," it makes an RNA with a sequence that is the same as one strand of the DNA, including the exons and introns. "Splicing" joins the sequences of the exons in the RNA together to form an mRNA by removing the intron(s). These are natural processes. "Reverse transcription" is a process that can be used in vitro to convert the mRNA to a DNA that does not exist in nature. But the sequence of this cDNA is the same as what would be obtained simply by cutting the exons out of the gene and joining them together. Why is a sequence of exons patentable but the sequence of the gene is not?

discovery should not in itself impinge on further discovery. But it can have an effect both before a patent is granted (if it affects the timing or conditions of publication) and after granting (in unusual cases in which the patent is enforced in a way that impedes scientific discovery).[38] A major effect of patents on science might be the spur they provide for directing research, in the sense that their potential availability could affect the choice of which projects to undertake.

Certainly, the wish to obtain a patent can affect the decision on whether work is ready to publish, and the need to apply for the patent before publication (in Europe) can pose a dilemma between the wish to obtain scientific priority and the need to protect the patent. But is this any different from the effect of competition between laboratories racing to get priority for a basic discovery?

It's not surprising that inventions or applications should often be duplicated—they are likely to occur when basic knowledge reaches a point that makes them possible—but why should basic discoveries in science often be duplicated? You might argue that this is a consequence of competition among the increased number of scientists today, but a list drawn up in 1922 of discoveries from 1500 to 1900 had almost 150 simultaneous discoveries (usually involving two claimants, but sometimes up to five), mostly in basic sciences.[39] In the past 50 years, more than half of the Nobel Prizes for Physiology or Medicine have recognized simultaneous, or at least related, discoveries. Perhaps paradigm shifts are not so much the result of unique inspiration, but happen because the load on a paradigm simply becomes too great?

If science were truly impartial, it would be irrelevant who made any particular discovery. Instead, a discovery might be regarded as an inevitable consequence of existing knowledge, rather than bringing acclaim to the individual scientist(s). The frequent occurrence of simultaneous discovery, when multiple scientists are working on the same problem because they recognize its interest, argues that discoveries are made when their time has come.

But science is a human activity. Scientists get kudos (in the language of sociology, rewards) for making discoveries. As a result, competition has been reflected in disputes about priority for as long as there has been science: Galileo disputed with Grassi about comets, Newton fought with Hooke about gravity and with Leibnitz about calculus, Donald Johanson fought with Richard Leakey about Lucy, the (potential) missing link, and Bob Gallo had a fierce fight with Luc Montagnier about the discovery of HIV (which eventually involved the governments of the United States and France). Ironically, Alfred Nobel had a dispute with Frederick Abel and James Dewar about the invention of cordite (he sued them for infringing his patent). These fights have tended to impede rather than advance science.

Competition is an inevitable companion to the obsessive character that drives scientists; until it gets out of hand, it is a spur to science, possibly even an essential part of science, whether driven by patents or by publication of research articles. Irrespective of the means used to present the data, the end is the same: an addition to knowledge. As a human activity, science can involve arguments and emotions on top of competition. These can affect the view as to what is reality. But even if the recognition of reality is delayed, sometimes long delayed, at the end of the day, data rule. That is the distinction of science.

CHAPTER
4

BIOTECH

I always maintain that the best attribute we had was our naïveté … if we had known about all the problems we were going to encounter, we would have thought twice about starting.[1]   Herb Boyer, 2001

Until the advent of biotechnology, there were only two modes for research in biology: "pure" science and "applied" science. Pure science was more or less the exclusive prerogative of academic institutions. Applied science was carried out by commercial companies, such as pharmaceutical houses, and was focused on practical developments rather than the pursuit of knowledge. The disdainful view that academics took of applied research was captured by the view that "research at Big Pharma consists of figuring out how to steal the results of other drug houses," circular though this might be.

Disdain for commercial activities was unique to biology: physics and chemistry had established a tradition of partnerships between academia and private companies. The trigger for following suit in biology was the possibilities opened by the creation of recombinant DNA. This led to the founding of several companies that functioned in a third way, performing (and publishing) basic research that would lead to commercial applications (which they would patent).

An early precedent for the biotech industry in relying on patents came from the first biological molecule to be patented, the hormone adrenaline. The function of the adrenal glands in increasing blood pressure was discovered in the second half of the nineteenth century. The active component, adrenaline, was purified in 1901 by Japanese chemist Jokichi Takamine, who established his own laboratory in New York but licensed exclusive rights to drug company Parke, Davis & Co. After the hormone was patented, as adrenalin, endocrinology took off as a field in which researchers isolated hormones from specific glands, and drug companies made them available for medical use.

Sometimes academic researchers would consult for companies and sometimes a company might support research at a university, but there was a division between the people involved in research (conducted in academia) and the process of scaling up and production (conducted by drug firms). Prestige accrued from isolating active biological compounds, such as thyroxine (thyroid hormone), insulin, or vitamin B12.

Researchers could patent their discoveries, but there was something of a feeling that they should not profit personally, and royalties often went to the university. Something of a backlash, however, led to a situation in which most universities became cautious about such collaborations. A pharmaceutical company might gain prestige from recruiting an academic researcher, but it would be unusual for a scientist to make a transition from a drug firm back to a university.

Some pharmaceutical companies set up their own research facilities,[2] but these focused more on applied research. Biology, in which basic research, such as in endocrinology, stayed mostly in academic institutions, was different from chemistry and physics, in which industrial research led to several Nobel Prizes (including research at IG Farben, General Electric, and Bell Labs) between the two world wars. (Researchers at Bell Labs were awarded five more Nobel Prizes between 1956 and 2018.) The only one awarded in biology outside of universities and research institutions was to Kary Mullis, who shared a Nobel in 1993 for inventing the polymerase chain reaction (PCR) at Cetus.

All this changed when the biotech industry took off in the 1970s. The same researchers were involved in academia and industry, they could move freely between the two, and the lines between basic and applied research became blurred. Researchers moved from being solely involved in basic research into becoming entrepreneurs who took advantage of their own discoveries.[3]

Founded in 1971, Cetus was one of the very first biotech companies spawned by molecular biology, although it did not move into recombinant DNA until later. Genentech, founded in 1976, is generally reckoned to have been the first company founded specifically to take advantage of recombinant DNA technology. Founded by a group of scientists in 1978, Biogen was also instrumental in establishing the model for close collaboration between the company and the labs of its scientific founders. A myriad of others followed suit, sometimes founded by scientists, sometimes by venture capitalists who recruited scientists. Today there are more than 6000 companies in the United States alone that describe themselves as involved in biotech. This is a significant change in the scientific landscape.

The scheme for founding Genentech was a direct consequence of Stan Cohen and Herb Boyer's creation of recombinant DNA (previous chapter). Bob Swanson had been working at Cetus, but had been fired after he had failed to persuade the company to move into recombinant DNA. He persuaded Boyer to join him in starting a company in 1976, and they approached the venture capitalists Kleiner and Perkins (who had funded Cetus) for the start-up funds. "Kleiner & Perkins recognize that an investment in Genentech is highly speculative but we are in the business of making highly speculative investments," Thomas Perkins wrote at the time.[4]

The plan was to take advantage of the patent that Stanford had filed on the Cohen–Boyer technique, with the objective of producing human insulin in bacteria. Animals were the only source of insulin at the time; it had to be obtained from pigs and cows and, aside from the expense of production, had the disadvantage that it could provoke immune reactions. Producing insulin from recombinant DNA was venturing into unknown territory, however, as to date no human protein had been produced in bacteria.

Boyer decided to start by getting proof of concept with a simpler protein, somatostatin, only 14 amino acids long, and started a collaboration with Arthur Riggs at the City of Hope, in Duarte near Los Angeles. They applied for a grant from the National Institutes of Health (NIH). It was turned down as being an academic exercise without practical merit.[5] So Genentech, which did not yet have its own facilities, funded the research, agreeing that the work could be published in the usual way, although with a prior agreement with the City of Hope that Genentech should have an exclusive license for any patents.

Their paper reporting the successful production of somatostatin in *Escherichia coli* was published in *Science* at the end of 1977.[6] Keiichi Itakura in the Riggs laboratory was the first author and Herb Boyer was the last author, a conventional enough order. Illustrating the differences between the worlds of patents and science, however, Boyer was not included in the patent on the grounds that he had not directly contributed to the experiments.

Whereas the tradition in science is to include everyone with a connection to the work as an author, the rule for patents is to restrict the list to people who can be demonstrated to have been directly involved at the bench. (Ironically, this may mean that a young scientist may get as much, or more, kudos from being named on a patent application than being one of the middle authors on a research paper.)

Publicity claimed this was the start of a new era, but the response was two-edged. There was a view in universities that it was a conflict of interest for a researcher working in a university laboratory to have two hats—that is, for his work to benefit a private company commercially while he was also working at the university. The traditional "purist" view of biology was dominant at the time—but many of the critics went on to form their own biotech companies within the next decade. This was a great change in the atmosphere in biology.

Collaborations with industry were not unknown, of course, but running a laboratory at the university while simultaneously being intimately involved in your own company was a new departure. Biotech companies

The Genentech campus in South San Francisco. By the time Genentech was sold to Roche in 2009, it was worth $100 billion.

started almost as offshoots of the university laboratory, attacking the same problems, sometimes with the same researchers continuing their work while making a transition from one to the other. Now the model has now become widely accepted.

At Genentech, somatostatin did not become a commercial product, but its production triggered a race to synthesize insulin by recombinant means. Genentech was nothing more than an office, staffed by Bob Swanson and a part-time secretary at the time. Despite this, Swanson was able to get a major drug company, Lilly, to sign an agreement to support research on insulin: Lilly was the largest producer of insulin from animal sources and had a pressing need to protect its dominant market position.

Although the definition of patent law concerning human genes was more than 30 years in the future, the approach followed by Genentech's collaborators, Keiichi Itakura and Arthur Riggs, was not to try to clone the human gene, but rather to make a synthetic DNA with the right sequence to code for it. This enabled them to design a sequence that would be optimally expressed in bacteria. (This followed the approach used in the somatostatin trial run, but was more complicated because of the larger size of insulin.) Pieces of this sequence were shipped to a rudimentary facility that Genentech established in South San Francisco to be joined together and cloned in bacteria.

Their competition was formidable: Wally Gilbert at Harvard was trying to clone insulin using a cDNA approach. The model for research was similar to Genentech's: the work was done in the Gilbert laboratory at Harvard, but was funded by the company Biogen, which Gilbert and four other scientists had established in 1978. It built a facility in Cambridge, Massachusetts in 1982. Gilbert described the effect on the relationship between the university and industry: "What we are seeing here is an attempt by academics to control the industrial development."[7]

In some other fields, notably computing, senior scientists sometimes left their academic positions to run companies. This was less common in biology, in which they tended more to drive projects at a company intellectually, while remaining at the university.[8] Junior scientists were more likely to move to a company on a full-time basis.

Senior scientists rarely do much research themselves: they spend their time directing postdocs and graduate students and writing grant applications. An association with a company that funds some of their research can influence the direction of their laboratory. This is not a new situation; scientists have always consulted for industry, and undertaken work under sponsorship.

The relationship can be accentuated, however, when a scientist has an entrepreneurial interest in a company. It can create a feeling that the postdocs and graduate students are being taken advantage of for commercial purposes, but this is not entirely one-sided: scientists who are known to have strong commercial associations may be better able to attract graduate students who believe this offers better long-term prospects for employment.[9] This is part of the development of molecular biology, from a young field devoted exclusively to "pure" research, to a mature field with commercial applications.

The model for running a project in tandem at a university laboratory and a company presented no conflict for university administrators: the outside company funded research, providing an alternative source of funds to the NIH or other public grants, and if patents were obtained, the university owned at least a share and would benefit from licensing revenue. The consequences for the scientific community were not so evidently one-sided. True, working directly for a company meant a reliable source of funding for a researcher, free of the need to apply for grants— but you had to be interested in the projects the company was focused on, and give up freedom to go off on your own tangent.

Yet one of the reasons the model worked was that the companies, recently founded by scientists well entrenched in academic life, acknowledged the importance of allowing their scientists to publish in research journals. Biotech's model of patent and publish contrasts with the reliance on secrecy often followed by Big Pharma, and was crucial in attracting talent; scientists could establish reputations much as they would in an academic setting, and indeed move freely between academia and biotech. This was helped by retaining close ties with the academic departments the researchers had come from. The atmosphere changed 180° from regarding industry with suspicion. One observer called the style of management in the early days at Genentech a "scientists' republic."[10]

Some of the early researchers at Genentech later moved back into academic positions in academic institutions.

Genentech won the race to synthesize human insulin.[11] One reason was that, as a private company, it was not subject to the rules about recombinant DNA research that applied to academic institutions (such as Gilbert's laboratory at Harvard). In fact, one of the reasons why Genentech set up in South San Francisco was that they rejected Berkeley because local ordinances had been passed there to ban work on recombinant DNA (see Chapter 11). (Although using a synthetic sequence rather than cloned material bypassed most of the problems.)

Doubting whether genes as such could be patented (this was farsighted), Genentech applied for a patent on the method of synthesizing and cloning the gene—one so broad that it might in fact be taken to cover any synthesis of a human gene. It was granted in 1982, but was controversial and became disputed, generating lawsuits in which hundreds of millions of dollars were at stake.

Synthesizing the gene de novo was a remarkable technical feat, but by its very nature did not reveal anything new about the biology. Cloning genes with medical importance, whether by synthesis or (more often) through making a recombinant DNA from a natural product, became an objective in its own right. Papers were published in leading journals. The lines between basic and applied research had blurred.

That blur in focus was one of the reasons the model worked. A young researcher could undertake a project that would both merit publication in a high-impact journal and have commercial significance for the company. This was partly the result of the way patent law was interpreted, in effect to allow cloning of natural products to be patentable. However, the patents were generally restricted to exploitation of only the natural product—that is, they excluded any subsequent development of variants with enhanced or better directed activity.[12] Those had to be the subject of separate patents, but although they could be highly significant for the company, they would not necessarily confer the same intellectual prestige. As companies moved into developing second-generation products, basic and applied research separated once again, and the attraction of working at a company faded if you wanted to acquire an academic reputation.

In the early phase of biotech, while the targets for cloning were natural products, patent and publish was a powerful model. Once the patent application had been submitted, publishing wasn't merely altruistic or designed solely to keep researchers happy: it prevented anyone else from applying for a patent. This was a contrast with the old model of Big Pharma in which secrecy was important to prevent work-arounds that could bypass a patent by making minimal modifications to synthetic structures.

Genentech's success, followed soon by similar feats at Biogen and other companies, changed the model for research in biology from pure academic to a hybrid, admitting companies as legitimate players in the intellectual arena. With increasing success of biotech leading to unimaginable growth for the companies, often resulting in takeover by a denizen of Big Pharma, management became far removed from the original founding scientists. This reinforced the move back to the original model for dividing science between pure and applied research. But it remains true that it's no longer necessarily a one-way street from basic to applied research.

Competition was the rule rather than the exception among the early biotech companies, perhaps because the most lucrative targets for recombinant DNA technology were so obvious. (By 1983, at least 26 biotech firms were trying to develop products based on interferon, a protein thought to have a powerful antiviral effect.[13]) The rush to publish soon turned into lawsuits about patents between companies that could be bitter and protracted; sometimes universities involved in the patents were drawn in.

Sometimes the clash between patent claims was direct, turning on matters of priority; sometimes it was indirect and might come from unexpected quarters. A patent for cloning a DNA might be challenged, for example, because it relied on using a protein that was a subject of another patent. (Protecting everything in sight with patents was a different model from the tradition in Big Pharma, in which competition more often took the form of developing drugs that were just different enough to avoid patent fights.[14])

It would be fair to say that biotech firms served the public interest no better than Big Pharma as they commercialized their first products. These were not necessarily significantly better than those previously available, and the prices were not necessarily lower.[15,16] Commercial attitudes were almost the antithesis of the value system of science. To the extent that scientists were implicated, this involved more of a split personality than simply wearing two hats.

Competition and lawsuits about patents could spill over into the academic sphere. Scientists became more cautious about sharing reagents. There had been a general agreement that reagents such as cell lines should be available on request, but if you received one, you would not pass it on to a third party. Having your reagents used for commercial purposes, or getting caught up in a lawsuit because a company had used a line in a patent application, discouraged free exchange without secure agreements on further use.

A sense of obsession is common among scientists, and can easily get out of hand. Why else would people be prepared to go through periods as PhD candidates and postdocs in which they hardly ever leave the laboratory except to sleep and eat? Those who have other interests and are less devoted are at a distinct disadvantage. This is not at all understood in the outside world.

I remember a meeting at the Banbury Center of Cold Spring Harbor around 1987 where scientists were matched with congressional staffers, many from the Dingell committee that was investigating scientific (mis)conduct at the time. This was not a pleasant occasion: the staffers showed an almost universal sense of disbelief that people could work so obsessively in the laboratory simply because they were interested in science.

Their attitude that scientists must be venal says more about politics than science. The only appropriate word I can find to describe their attitude is "vicious." They kept probing to figure out what sort of longer-term remuneration the scientists thought they would get from this; this was a witch hunt in which they felt there had to be a hidden reward, and simply could not believe there was none. If you don't understand this level of obsession in scientists, you will never understand science.

A sense of being different, or at least a strong sense of individuality, is common among scientists. I once asked a group of scientists, "Did you go into science because you felt alienated from society, or at least had some sense of discordance, or did you become alienated as a result of going into science?" and the answer was a universal chorus that a sense of difference had been a driving force for entering science. I am not sure that is still so true: especially under the impact of companies providing an alternative to academia for research, these days I get more of a sense that science is a career like any other and not something so distinctively different. If scientists lose that obsessive drive, the atmosphere of science may well become different.

The naïveté of researchers about the difficulties of exploiting their discoveries for commercial purposes at the outset of biotech was matched by the impossibility of foreseeing the consequences for academic research. The relationship between biotech and academia is reciprocal. The origins of biotech in academic research had a major influence on the way the industry was at first constructed. Biotech, in turn, has had as much impact on the conduct of academic research. It has changed the scene by ranking practical objectives equally with new intellectual insights, with knock-on effects on the direction of research. "Applied" research has become more respectable. It has made patentability a legitimate objective in academic research, which represents something of a reversal in opinion in biology. It has opened new career paths, with indirect effects on the choice of research topics. The interactions between academia and biotech are less intense than they were at its creation, but they are long-lasting.

PUBLISHING

CHAPTER 5

THE MYTH OF THE SCIENTIFIC PAPER

A scientist is supposed to have a complete and thorough knowledge, at first hand, of some subjects, and, therefore, is usually expected not to write on any topic of which he is not a master.[1]

Erwin Schrödinger, 1944

A scientific paper is a variation on the theme: history is written by the victors. It is supposed to be a logical construct based on the data. But this is a myth. More often, the principle of the paper is really *post hoc ergo propter hoc* (a fallacy arguing that because one event follows another, the first event must have caused the second). Essentially the argument of the paper is constructed retrospectively to create a logical framework for the results (as opposed to whatever the authors were actually thinking when they started the work). It's how and why the authors would have done the experiments, if they had known at the beginning what they found out at the end.

Most scientists write their papers according to the prescribed framework, but beyond that, they are not very interested in theorizing about the process of science. Sir Peter Medawar, awarded a Nobel Prize in 1960 for his work in immunology, was an exception who wrote about the process of scientific discovery. He once gave a talk on the BBC entitled, "Is the Scientific Paper a Fraud?"[2] He was quick to explain that he did not mean it misrepresents facts or conclusions, but that it misrepresents the process of discovery.

Medawar argued that people had confused the role of formulating new ideas as opposed to testing them. "Hypotheses arise by guesswork.... .

I should say rather that they arise by inspiration; but in any event they arise by processes that ... [are] ... certainly not of logic." So whatever the process of designing experiments, it is not as represented in a paper. "The scientific paper in its orthodox form does embody a totally mistaken concept, even a travesty, of the nature of scientific thought," he concluded.

One of the most impressive scientific papers I have ever read was François Jacob and Jacques Monod's proposal in 1961 for the way genes are controlled in bacteria.[3] It has an ineluctable logic, argument following rigorously from argument, that could only have originated in the tradition of French philosophy. Jacob and Monod shared a Nobel Prize in 1965. But this is how Jacob described the process of authorship in his autobiography: "Writing a paper is to substitute order for the disorder and agitation that animate life in the laboratory ... to replace the real order of events and discoveries by what appears as the logical order, the one that should have been followed if the conclusions were known from the start."[4]

Support for the view that scientists do not write up their results in a way that reflects the actual process of discovery comes from a survey showing that the vast majority of published papers claim they have successfully tested a hypothesis. The authors of the survey thought this indicated increasing reluctance to publish negative results.[5] But scientists would be awfully perspicacious if their work usually correctly anticipated the results.

A more likely explanation is that in most cases, the "hypothesis" was in fact conceived after the results had been obtained. This is equivalent to the phenomenon sometimes called HARKing by sociologists, an acronym for hypothesizing after the results are known.[6] (However, scientists certainly don't like to publish negative results: the *Journal of Negative Results in Biomedicine* was founded in 2002 and published only 224 papers before it folded in 2017.)

Outsiders both underestimate and overestimate the significance of negative results in basic science. People do not realize that much scientific effort leads nowhere and is never published. Blind alleys are inevitable. One recent book argued that it is a great disservice for all that negative data to stay in laboratory notebooks instead of being available to other scientists.[7] But that misunderstands how science really works. Perhaps occasionally it would save another scientist from following a blind alley, but for the most part, it would be ignored even if it were published.[8] And publication of negative results would enormously clutter up the scientific literature. Consider a *reductio ad absurdum*. Suppose you conceive a theory that the earth is really flat. You test it and find out you are wrong. Will people be interested to read your analysis?

A more formal way of looking at it is that the reward system of science gives acclaim to scientists for obtaining positive results, but not for negative results. The one exception is when negative results disprove a long-standing, accepted theory. Of course, obtaining negative results is also an important aspect of the self-correcting mechanism of science—but although this may be important for science as a whole, it does little for the scientists making the report. Along the same lines, this is why although a hypothesis can be regarded as valid only if it is in principle falsifiable, in practice scientists try to find support for a hypothesis, rather than try to disprove it.

I don't know whether scientists themselves are under a conscious or subconscious illusion that they proceed from testing hypotheses, to obtaining data, to writing up papers, but the scientific enterprise has done a fantastic job of maintaining its mystique. Perhaps that is partly why there was such an uproar when Jim Watson wrote his account of the race to find the structure of DNA. Many of his contemporaries objected to his memoir, *The Double Helix*, some on personal grounds, some for more general reasons. Maurice Wilkins objected that it presented "a distorted and unfavorable image of scientists."[9]

Attempts to stop publication resulted in Harvard University Press dropping the book, even though Watson was a professor at Harvard at the time. The book became a sensation when it was published in 1968, not least for the way it personalized how science works. It probably should be taken no more literally than the idealized view of science, which seems to have survived anyway, not least in the attitude to the research article.

Scientific papers were often written in the third person to give an impression of impartiality. This was downright silly, because you can't escape the fact that people conceived and executed the experiments.

At least today, papers are more often written in the first person with regard to how experiments were performed, even if the description of the results is surrounded by a mythic aura of logic.

I once put the view to a *Cell* author that research papers are exercises in spin that should be judged on the data, ignoring the introduction and discussion. He took me at my word, and sent me a set of six figures showing data, with minimal explanation of how the data had been obtained. He then asked me to tell him whether I thought this would be a suitable submission to *Cell*. It was rather difficult to work out, and it would probably have been almost as difficult to assess a paper that actually reported experiments in the order that researchers had performed them, with an explanation of why they had really done them. So even though the scientific paper shows something of a "one-size-fits-all" approach, it serves a useful purpose in presenting information in a form that is readily assimilated.

A scientific paper has a very stylized, you might almost say stereotyped, format. There are essentially four parts to a research paper: the Introduction is where the retrospective logic is important; the Results presents the supporting data; the Discussion explains the authors' view of its significance (puts their spin on it, you might say); and the Materials and Methods or Experimental Procedures (various names are used) gives sufficient details for any interested researcher to repeat the work.

The heart of a paper, of course, lies in the Results and the Materials and Methods. Results should give a representative set of data; representative means that the experiments should have been repeated enough times to make clear what is typical, and what might be rejected as an outlier. It's impossible to quantitate, but I suspect that most cases of papers in which the results later prove to be incorrect arise because the data were not truly representative (sometimes because the experiments were not repeated enough times).

Peer review is the glue that holds the scientific process together. It is the first line of defense, before we get to questions of reproducibility or to self-correcting mechanisms. A journal usually sends a submitted paper to two reviewers, sometimes to more reviewers for interdisciplinary work in which multiple types of expertise are needed. A peer reviewer should be someone who is knowledgeable about the field and

Myth	Reality
Introduction	
account of prior work and logic for performing experiment	reconstructs rationale for experiments in view of actual results
Results	
the data, the whole data, and nothing but the data	representative data to support the conclusions
Discussion	
explains significance relevant to prior work and future implications	the authors' view of the results and claim for priority
Materials & Methods	
all the details needed to repeat the work	those details that were recorded as the work was done (or sometimes retrospectively)

A view of the myth and reality of how scientific papers are written.

familiar with the techniques that are involved, but who does not have any direct interest in the results, and certainly is not a competitor.

Peer review started in the nineteenth century as part of editorial consideration: in the period before specialization, the Editor of the journal (or later an editorial committee) would in effect review a submitted paper. Reviews by anonymous referees were introduced by the Royal Society in 1833, but this became sufficiently controversial that when a committee of the Geological Society of London assessed the system in 1903, it considered banning the term "referee."[10] Complaints were basically the same as today: the system could be biased or capricious, and might deprive researchers of their priority.

The system became common in the Anglo-Saxon world, but not in Europe: Albert Einstein was famously outraged when a paper he submitted to the American journal *Physical Review* in 1936 was sent out for review. "We had sent you our manuscript for publication and had not authorized you to show it to specialists before it is printed. I see no reason to address the in any case erroneous comments of your anonymous expert. On the basis of this incident I prefer to publish the paper elsewhere," Einstein wrote.[11] But when Einstein published the paper elsewhere the following year, its conclusions had been significantly changed, in fact reversed to demonstrate instead of denying the existence of gravitational waves. This accorded with the reviewer's criticisms![12]

External peer review became common only in the second part of the twentieth century. "Peer review" replaced the concept of the "referee" when government took over financial support of science after 1945, and needed to justify how it decided on funding.[13] The journal *Science* started external peer review as a matter of routine in the late 1950s;[14] by the time I arrived at *Nature* in 1970, papers might be rejected without external review, but would be accepted only after peer review.[15]

A submission to a journal is supposed to be a privileged communication: no one who sees it should take any advantage from having seen it. Scientists take various views if they are sent a paper that appears to be competitive with their own work. The most rigorous and honorable response is to return it without looking beyond the title page. But, of course, there is a temptation to look inside and see how far the competition has got.

"While there is often the feeling that referees should refrain from taking advantage of their position ... only an insensitive fool could let his students go on with an experiment when his 'insider' information tells him that it no longer makes sense," Jim Watson says.[16] But there is a difference between stopping a student from following a blind alley and changing your line of research or expediting your own publication. "It is not only second-rate scientists who behave improperly, but often those of the first rank," Watson adds.

He goes on to say, "If you are unknown and with little resources at your disposal, you may be certain that if you talk openly, other labs will rush in, and you may quickly be frozen out of a field that owes its existence to your ingenuity. The only sure way to prevent such a disaster is to let no one outside your lab onto your secret until you have exploited all its obvious consequence."[17]

But once you have reached the point of submitting a paper to a journal, you have not necessarily reached safety. One Nobel Prize winner said to me quite freely that he regards it as fair game to go after a topic if he sees a copy of a submitted paper. That's a minority attitude, but it's relatively common to write up your own work immediately, or to pressure a journal to publish it more rapidly, if you learn about submission of a competitive paper. (I discuss some recent famous—or infamous—examples in Chapter 18.)

Because priority is important, research articles carry a received date, showing when the article arrived at the journal, and, if the article went through a cycle(s) of revision, a revised date showing when the last version was submitted. (This also lets readers assess the efficiency of the journal.) It is not unknown for authors to try to rewrite a paper on the page proofs if they learn of other results that affect their interpretation. The rule at *Cell* was that changes after acceptance should never affect the substance of the paper, in fact they should be absolutely minimal. Of course, now that papers are published online more rapidly after acceptance, opportunities to abuse the system are reduced.

Abuse of the process to the extent of delaying review of a paper or giving an unfair review is not common, but it's not unknown. It's impossible to quantitate (especially as you don't always know who is working on a particular topic), but it did not seem this was a substantial problem in external reviews of papers submitted to *Cell*. In the majority

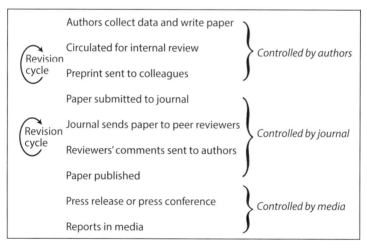

Filters en route to publication and publicity. Authors control the process of writing, consulting, and revising the paper until it is submitted to a journal. The journal controls the process of publication by choosing peer reviewers and setting conditions for publication, which may include revisions of data and presentation. Once the paper is published, the contents of a press release or press conference may influence reports in the media, and it is at this stage that exaggerations may creep in.

of cases, the reviewers were in general agreement on a paper; in a significant minority, there were disagreements that had to be ironed out; and for only a small number (<10%), would the differences be irreconcilable. (Such differences were usually handled by putting the issue to another reviewer.) In that last category, there would sometimes be a case in which it seemed that a reviewer's opinion was due to something other than fair criticism of the data.

Attitudes vary on how to treat someone who comes second in a competitive situation. *Nature* at one time had a rule that it would not consider a paper reporting the same observation as one that was already in press. An editor refused to consider Luc Montagnier's report of the sequence of HIV (the virus responsible for AIDS), because Bob Gallo's report of the sequence had already been accepted. Such a policy really puts the screws on, and in my view goes too far. The rule at *Cell* was that if a second (or third!) paper was submitted within the same publication cycle as one in press, we would try to publish them together with the first one if they were accepted.

At the other extreme is the fate of Diane Pennica's paper on the cloning of tPA (plasminogen activator) at Genentech in 1982. After she had cloned the gene, but before she had written up the work, she received a call from a scientist who said that he had been sent a paper for review reporting the cloning of the gene. He asked if she had cloned it and had a paper in press, in which case he would recommend rejection.[18] This was surely disgraceful: an abuse of confidence in telling a competitor about a privileged submission, and turning science into a game of "winner takes all" (without even knowing, after all, if the sequences were the same).

"I think peer review is hindering science. In fact, I think it has become a completely corrupt system," Sydney Brenner said, in a conversation recollecting his career at the Laboratory of Molecular Biology in Cambridge.[19] This is an extreme view, but it is certainly true there is no metric to measure the efficacy of peer review of submissions. There's a widespread view that it's an essential protection, not only for the body of science, but to help authors to avoid publishing mistakes that will damage their reputations.

Peer review is admittedly an imprecise process, but it is better than any alternative. As Churchill said about democracy, "it is the worst form of government except for all those other forms that have been tried from time to time." The strength of peer review rests on the alternation of roles between authors and reviewers, so everyone has an interest in making it work. Its weakness is that reviewers can only review on the basis of the data presented by the authors, and furthermore reviewers and authors in a field may share the same implicit (sometimes misleading) assumptions.

Reviewers are usually anonymous, so that they have a free hand to criticize without fear of reprisal. Some journals have now changed to signed reviews, but this is a small minority. I used to regard it as axiomatic that reviews should be anonymous at *Cell*, but I am inclined now to the view that if your criticisms are fairly based, why shouldn't you admit to them? I suspect the price of signed reviews may be that criticism is watered down, or that researchers decline to review papers when they feel very critical, but the advantage is that it would eliminate most abuse of the system.

Of course, abuse can be a two-way process. An author might respond to an adverse review—even a fair one—by retaliating on another occasion. I also worried at *Cell* that authors might try to nobble the reviewers if they knew who they were.

It's sometimes proposed that reviewing should be double-blind: the reviewers should not be told the names of the authors. This is hopelessly impractical. One of the first things you do in reviewing a paper is to look at the previous work to see if it's a solid foundation for the new paper. It would be a terribly incompetent reviewer who could not deduce from previous work at least some of the laboratories involved in the new submission (or alternatively that the laboratory was a complete newcomer to the field).

Anonymous or not, criticism in science used to be fairly direct. Seminars or talks at meetings could come under fierce attack if the data were judged inadequate. That has changed. An early indication of how the climate has turned against any real criticism came from a Cold Spring Harbor Symposium, when there was a session on the structure of the recently discovered nucleosome—the basic building unit in which DNA is packaged by proteins. There were some pretty dire talks with more speculation than data. After one of them, Francis Crick said with some irritation, "You amateurs should keep out of this field and leave the model building to us professionals."

Perhaps it was due to the *ad hominem* nature of the comment, but the audience booed Crick (who then absented himself from the meeting for the next three days). But it is rare now to hear real criticism, even of data that are evidently incorrect or misinterpreted, whereas you used to present incomplete data at your peril. (The speaker at Cold Spring Harbor consoled himself by going around saying to people, "Well, at least it was me Crick picked out to criticize.")

I believe in intellectual brutality, by which I mean applying unsparing rigor to the data. Our attitude to submissions at *Cell* was that we would publish them only if we could not find the flaw. Most of the mistakes we made were due either to failure to apply sufficient rigor or to being blinded by the common assumptions of the field. It would be a pity, in fact I would maintain it loses the essence of science, to abandon this attitude in order to bring a kinder and gentler approach to the individuals involved.

You are not always aware of your assumptions. I remember a paper submitted to *Cell* reporting a discrepancy between the sequence of a protein and the sequence of a gene. It was impossible to believe the result, but reviews by a series of experts in all the techniques that the paper used did not turn up any flaw. So we felt compelled to publish the paper, but it was rapidly proved to be incorrect. That's the price you pay for judging a paper on the data and nothing but the data: but it's worth paying. Fortunately, this did not happen too often.

When we reviewed the paper, one perceptive reviewer asked how many times the work had been reproduced. The authors answered seven. We felt we could hardly argue with this. However, we assumed that the answer meant that the work had been repeated completely from scratch seven times. It turned out that in fact a single protein preparation had been made at the beginning, and then the work had been repeated seven times using this preparation. An error had occurred in the original protein preparation, so the subsequent seven repeats were irrelevant. It's the assumptions that kill you: we never thought to ask whether the seven repeat experiments always started from the very beginning.

Authorship might seem to be the least formalized part of a scientific paper—what can it be but a list of people involved in the work?—but the order of authors can be a contentious issue. The convention is that the first author is the person who did most of the work, and the last author is the group leader, the person who provides the intellectual stimulus and obtains the funds to support the work. Positions in between on the list can mean anything, from significant contributions to a sort of honorary authorship simply for providing reagents. When you are a junior scientist who will soon be looking for your next position, it's crucial to be first author on at least one paper.

It's the role of the senior author to resolve any disputes and decide the list of authors. Usually, the issue is settled within the laboratory before the paper is submitted and is fait accompli to the outside world. Occasionally, it reaches the journal when the senior author changes the order of authors after submission. Just once at *Cell*, I was asked to intervene, when an author who was in the middle of the list called me to argue that he should be the first author. I explained that I had no authority to interfere in what was essentially an internal matter for the laboratory.

He wanted us to pull the paper from the issue, which posed a more difficult question: what do you do if an author tells you he does not agree to have his name on a paper? And as a practical matter, the issue had already gone to press, so it was impossible to delete the paper. When I said there was little I could do in these circumstances, the author informed me he would get an injunction to stop publication.

Although this was something of an idle threat, it would not have been a comfortable situation to publish a paper whose authors were in open disagreement. In the end, the dispute was settled by keeping the order, but adding a footnote that the names were in alphabetical order. More often, two authors' names may carry an asterisk with a footnote, stating that they played equal roles in the work. Once a paper has more than two authors, recognition is important for those who aren't at the beginning or the end.

Authors could occasionally get a bit cross when their papers were rejected, but for the most part respected the process even if they did not agree with the decision. There are always other journals. The only time it turned personal was when we rejected a paper after an unfortunately long delay; admittedly we had not handled it well, and the author (who was under pressure because he was up for tenure) called me to say, "You are not fit to be the Editor of a major journal, you are a disgrace to the scientific community, and I am going to get your Ph.D. revoked." Fortunately, I do not believe the University of Cambridge has a mechanism to do this… I met the author a year or so later at a scientific meeting, and he turned out to be quite amiable: I concluded his anger was a demonstration of the effect of the pressure of the tenure process.

Even though papers are supposed to be judged on the data they present, the authors' past record inevitably affects the reviewers' judgment. Authors with a good track record have an easier time than authors whose work has previously been called into question. Although this is rarely reflected in overt criticism of the data that's presented, it can affect the reviewers' willingness to accept supporting observations that are mentioned but not shown.

Most papers contain references to other, related work in the authors' laboratory that is not included in the paper. Sometimes reviewers ask to see the data, especially in cases in which the authors have not established themselves. Scientists work hard to establishing a reputation for reliability, and it's not unreasonable this should be reflected in willingness to accept their word about supporting information.

I became more suspicious of unpublished observations after a reviewer asked to see the data in one case, and the response came from the authors: "The data are not good enough to show the reviewer." So our policy became that unpublished data had to be clearly cited as such in the main text; you could not include them in the reference list along with citations to peer-reviewed papers. Of course, with the move to online publication, this has now become irrelevant because supplemental data are routinely included with papers, and everyone can see them, as described in Chapter 8.

Scientists and journals have to come to grips with the fact that a scientific paper can no longer represent an end point. There's always been an issue of when data are complete enough to make a paper or whether further work is required before publication is appropriate, but it's a new question to ask whether a single paper, or even a series of papers, is an adequate format in which to present the new data.

"Conventional journals have evolved to publish papers that have a 'bottom line.' In a typical genomic exploration of the kind for which microarrays were designed, reducing the results to a few 'bottom lines' would be like representing human anatomy with a stick figure. Clearly, a fundamentally new kind of scientific publication is needed," David Botstein concluded in his call to abandon hypothesis-driven science.[20] What else could the new format be but databases on the Internet?

It has become common for research papers to carry a statement that the authors have no conflict of interest. This is not much more than a sop to exclude more extreme cases in which authors have had a commercial interest in exploiting their results. But doesn't it miss a beat? What about conflicts on the part of the journal? With most scientific journals now owned by publishing conglomerates that have extensive and varied commercial interests, what about their conflicts? Springer Nature, for example, who own the influential general science journal *Nature*, have interests in China; can one be sure that their commercial interests do not influence editorial decisions in their journals?

There were protests from the academic community in 2017 when Springer Nature openly censored the contents of some of its journals for distribution in China to appease the Chinese government.[21] "A small percentage of our content (less than 1 percent)" was limited in mainland China in order to "comply with specific local regulations," Springer said. The withdrawn material actually comprised more than 1000 articles in two political science journals.

I do not know whether the COVID virus originated by a laboratory leak or by a natural exchange from an animal source. No one does: and possibly, no one ever will. I do know, however, that science should consider both theories on the basis of the data, but there's a worrying trend for journal editors to take a one-sided attitude. Of course, this is scarcely an unknown situation; science is replete with examples in which a dogma based on what seems to be a substantial body of data has been used to deny a hearing to alternatives. But here there are really no compelling data. Papers speculating that an animal model is responsible seem to have a much easier time getting published than those arguing for a laboratory leak. This does not seem to be due to any difference in the quality of the data on either side of the argument. So what is responsible for this prejudice?

A recent Harvard graduate once said to me, "That thought is unthinkable." He meant that no right-thinking person would think it. That is not an acceptable position in science, which should function as an extension of Voltaire's principle: "I detest what you say, but I will defend to the death your right to say it." What matters is the logic of the argument based on the data, not whether you like the conclusion. If the conclusion is unacceptable, look for the flaws in the data and the logic.

The scientific attitude has remained in principle the same ever since experimental science began and has long been regarded as sacrosanct by scientists. But much else has changed in the modern era, with a move from individual practitioners to vast collaborative teams, limitations on presentation removed by the transition to publication via electronic media, and data being deposited directly in databases. It becomes reasonable to question whether the scientific paper should still be regarded as the principle means of communication, or whether this is outdated and we should look for other means of presenting results.

The ideal that scientists should only author papers on topics on which they are expert has been stood on its head: when a paper has tens of authors, it is unlikely that any single author has sufficient expertise to represent the whole paper. Authors are reduced to contributors on specific topics (on which admittedly they are expert). As the concept of the scientific paper transmogrifies into the electronic era, concepts of authorship and of what constitutes an appropriate contribution to science, may change further.

CHAPTER
6

REPRODUCIBILITY

The natural scientist is concerned with particular phenomena … he has to confine himself to what is reproducible… . I do not claim that the reproducible by itself is more important than the unique, but I maintain that what is essentially unique eludes treatment by scientific methods.[1]

Wolfgang Pauli, 1961

Publication is only the first part of contributing your data to science. The second part is ensuring that the work can be repeated. If there is any single concept that unifies all science, it is reproducibility or, to be more exact, the requirement that an observation can be reproduced by other researchers. Any observation that cannot be reproduced is in effect expelled from the canon of science. The need for reproducibility drives the way that science is practiced and reported.

Every researcher keeps a laboratory notebook with a record of reagents used and procedures followed. Internally in the laboratory, an experiment should be repeated a sufficient number of times to be sure that an exact way of obtaining the results can be reported. Externally, those details are presented in the research report. The key to the principle of reproducibility is that the experimental procedures should give all the details that are needed for an independent researcher to repeat the experiments. This is the opposite of a recipe in which the chef protects his originality by leaving out some crucial ingredient.

The search for reproducibility can be taken to extremes that would not be expected from reports of standard laboratory procedures. During the period when recombinant DNA research was difficult in the United States, as mentioned in Chapter 4, Wally Gilbert and his team at Harvard were trying to clone and express the insulin gene, but were impeded by local regulations. They went to the top-secret Porton Down facility in England, where there were extreme containment facilities. The materials they took with even included diluted reagents, in case diluting with the water at Porton Down might introduce any difference. "Different water can often kill you in the enzymatic reaction," Lydia Villa-Komaroff said,[2] a demonstration that any detail, no matter how small, may influence the results.

What happens when another laboratory is unable to confirm a published report? Of course, there is an intrinsic bias in the view that the self-correcting mechanism of science removes errors and that the frequency of incorrect reports is small. Attempts to reproduce experiments are usually made only when some discrepancy arises in later work. This has to be quite important to divert a researcher from following his own line of research (not to mention the fact that it is difficult to publish studies simply repeating other studies). The number of cases when discrepancies were noted but simply bypassed is unknown.

A survey by *Nature* in 2016 reported that a majority of respondents said they had tried and failed to reproduce another laboratory's experiments; and many had even failed to reproduce their own experiments![3] Of course, surveys are notoriously unreliable, and problems resulting from bias in responses make this anything but a definitive view of science. But the most interesting fact was that a majority of respondents also said they believed the majority of research reports was reproducible![4] Their skepticism was a lot less than you might have expected from their own experience. Is belief in the scientific system rational or more a matter of faith?

A rare direct investigation of published papers produced some disquieting results. The Reproducibility Project attempted to independently replicate key experiments from some of the most important papers in cancer biology, selected mostly from *Nature, Cell,* and *Science.* "The initial goal was to repeat 193 experiments from 53 high-impact papers published between 2010 and 2012, but the obstacles we encountered at every phase of the research lifecycle meant that we were only able to repeat 50 experiments from 23 papers," Tim Errington says.[5]

It is virtually impossible to describe research procedures so thoroughly as to give every detail an independent researcher needs in order to repeat the experiment exactly. The best you can hope for is that the missing details are not critical. So it is not entirely surprising that the project should have found a substantial proportion of cases in which they had to ask the original authors for further information.

It's more of a surprise that in more than one third of the cases, the original authors refused to help at all. Of course, when you are running a laboratory in a competitive field, you want to focus on current research, and it's a nuisance to be picked out for an academic study on reproducibility. The diversion of time and effort is not at all welcome, but too bad: that is part of science. Refusal to give information, or in some cases to share unique reagents, undercuts the whole basis of science. It's a weakness in the system that there are no repercussions when authors refuse to cooperate.

In those cases in which enough information was gained to try to replicate the experiment, less than half could be reproduced.[6] Earlier attempts from two drug houses, Bayer and Amgen, at reproducing preclinical studies in cancer found that even fewer studies could be adequately replicated;[7] whether this results from attempts to protect the work by secrecy, rushing to obtain priority, or simple sloppiness, it could be an explanation of why so many clinical trials ultimately fail (only 10%–20% lead to drug approval).[8] It should not be an open question whether the principle of reproducibility is honored more in the breach than the observance.

There's an irony in the relationship between preclinical studies and clinical trials. Research scientists should be the people who care most about reproducibility; people who conduct clinical trials care much more about validating their drug (as seen from many cases when negative results of trials were suppressed).[9] In the language of sociology, the reward system of science gives short-term benefits to researchers who publish interesting results, and only potential long-term disadvantages to those whose results might later be questioned as not fully reproducible. But for a drug company, it's crucial that a preclinical study should be reliable: otherwise, they risk wasting a small fortune on a clinical trial that may prove negative. This is a strange reversal of the incentive to ensure that research is reproducible.

Reproducibility is a widespread problem. It's particularly poor in empirical economics (in which data are collected or analyzed outside of an experimental framework) at <23%; it's higher, at ~64%, in experimental economics.[10] Social sciences don't do very well with only 62% reproducibility;[11] psychology is particularly dire at 36%.[12] Just as worrying, in all these fields, in those cases in which the results were reproducible in the sense of producing an effect in the same direction as the original report, its strength averaged only 50% of the original claim. This has been called a "reproducibility crisis."[13]

It's understandable, perhaps even predictable, that areas such as economics, social sciences, and psychology should have problems caused by difficulties in controlling conditions, but it is very disappointing that such problems should apply to biology. And the results of these attempts at reproduction are rather suggestive that authors everywhere are selecting their statistical tests to obtain the most positive results. Unfortunately, that's a practice as old as science itself.

Problems with reproducibility make it even more important that science should have a robust mechanism for correcting errors. Self-correction shows at its most dramatic when an observation is interesting but wrong. Take cold fusion. Stanley Pons and Martin Fleischmann at the University of Utah announced their discovery at a press conference in 1989. There was no scientific paper, and they were reluctant to reveal details of the experiment. In spite of the difficulties this created, there was a spate of attempts to replicate, and in short order, it became apparent that cold fusion was an embarrassing mistake.[14] The authors might have been spared much ridicule if they had submitted their paper for peer review before announcing the results. This was a vivid demonstration of the feedback mechanism that keeps science straight.

Some problems with reproducibility result from a bias in the way research is reported. It is not a bad idea to keep in mind the dictum "lies, damned lies, and statistics," when assessing research that relies upon reaching a statistically significant level to make its point. Remember that the traditional confidence level of 0.05 means in effect that the results have a 95% chance of being significant. The corollary is that they have a 5% chance of being a random coincidence.

Look at this way. Suppose 20 researchers perform the same experiment (without knowing about one another) and 19 of them obtain results that do not reach the 0.05 confidence level, but one does. The 19 researchers would almost certainly not bother to publish their results, but the one successful researcher would very likely do so. The results would be accepted as significant, but in due course repetitions would no doubt produce a consensus of insignificance, and the paper would have to be retracted.

This is an extreme model, but it points to the consequences that can result from the reluctance to publish negative results. Put another way, out of every 20 research papers that report just reaching an 0.05 confidence level, one is statistically likely not in fact to be significant. So a certain level of error is built into the system by statistical fallacy. The bias in publishing only positive results may help to explain cases in which papers are published that appear to be significant, but that cannot subsequently be repeated.

Reproducibility was not at all a contentious issue until the recent period in science. Fifty years ago, for example, most experiments used common ingredients. If a special apparatus was built, it was sufficient to describe its construction, and new technology was not difficult to understand.

The situation changed when unique ingredients began to be used as the result of cloning cells or DNA. When a paper reports an analysis that is based on a specific cloned cell line or a cloned DNA, the experiments cannot really be repeated without access to that material. Our policy at *Cell* was that publication required a commitment to make clones or any other unique materials available on request. My view remains that if a paper uses unique materials or data that the authors will not share, it is not a legitimate contribution to the scientific literature.

This can be a bit tough, of course: if you have spent time obtaining a cell or DNA clone, handing it out enables any competitors to catch up with you. But if the clone is unique, this is the only way to ensure that work is truly replicable. (In any case, anyone who did not wish to conform to our policy could publish in another journal.) I was very rarely called on to enforce the policy. Usually, it was sufficient for a researcher who was having difficulty obtaining materials to tell the other party that he had sent a copy of his request to *Cell*. I don't think I was asked to intervene more than once or twice in all the years I was Editor of *Cell*.

Science is supposed to be about the data, the whole data, and nothing but the data, but when research is criticized, either externally or even (perhaps especially) internally by a "whistleblower," there is an all too human tendency to circle the wagons and defend the research, rather than really investigate its integrity. I discuss some sad cases of fraud in Chapter 10. Retraction is viewed as a blemish on the record—a last resort—rather than a natural step in keeping the scientific record straight.

Of course, science is a human endeavor, and researchers can become committed to an idea. It is not unknown for all the energies of a laboratory to go into confirming data they have published that others have questioned, or for that matter to go into refuting a proposal from another laboratory with which there has been a conflict. Sometimes such arguments can continue for a long time. Hypothesis-driven science can become hypothesis-obsessed science.

I had mixed success in persuading authors they should publish a retraction in cases in which evidence had accumulated that a paper was wrong, but they had not volunteered a retraction. The usual excuse was that more time was needed to assess the situation. An honest answer came from one Swiss scientist who said, "Ho Ho, Benjamin, you do not expect me to say that I was *wrong*?" The paper was never retracted, even though the author himself had published contradictory data in another paper (without reconciling them with the earlier paper in *Cell*; one wonders what the reviewers of that paper were thinking.)

The research paper remains nominally the way in which results are communicated to the scientific community, but it is no longer necessarily the source of the primary data. If the human DNA sequence were written out in text the same size as this book, it would require 1,500,000 pages. Anyone who wants to consult the sequence goes to the database.

At a workshop in Bermuda in 1996, the participants in the Human Genome Project agreed that DNA sequences would be released into public databases within 24 hours after generation. The rules for rapid release became known as the Bermuda Principles. Indeed, this was one of the stumbling blocks in dealing with the proposal from Celera to complete the project (see Chapter 17), because, as a commercial company, Celera required at least some period of controlled access.

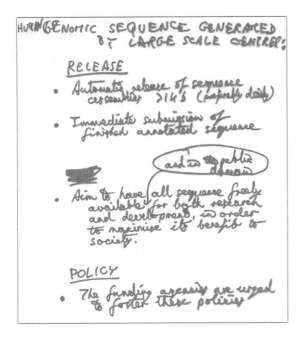

John Sulston's notes (he was in charge of the British sequencing effort) on a whiteboard at the meeting in 1996 led to the Bermuda Principles:[15]
- Automatic release of sequence assemblies >1 kb (preferably within 24 h).
- Immediate publication of finished annotated sequences.
- Aim to make the entire sequence freely available in the public domain.

There was an interesting breach in the policy of requiring sequences to be submitted to databases when Celera published its draft sequence of the human genome in *Science*. The Editor of *Science* made an exception to the journal's policy on the grounds that when publishing data from a private company, they needed to recognize the company's right to protect its investment.[16] My own view is that this was a most unfortunate breach in the principle that publication means information should be available to all.

Celera agreed to make the sequence available on its own website, but under restrictive conditions that in no way could be considered to satisfy any need for public access.[17] Because it was impossible to verify, was the paper essentially propaganda rather than a contribution to research? The sequence was not put into a public database until 2005, when revenues from licensing the information were falling sharply.[18]

There are now databases for protein sequences, DNA sequences, mutants in any number of organisms, genetic maps, and so on. In fact, most scientific journals require DNA and protein sequences reported in articles to be submitted to a public database. Regarding the database as the primary source (whether or not the sequence is part of a scientific paper) is a change from the traditional view that the process of peer review maintains the integrity of science when papers are submitted to journals. A research article can do no more than summarize interesting aspects of sequence data; scrutiny of the data itself requires the database. Has quality control passed from the journal to the database managers?

Indeed, in the case of protein structure, journals may now require a "validation report" from the PDB (the database for protein structures) as a condition of submission. Validation takes the form of a series of tests for common errors or inconsistencies.[19] Another sign that the database may be becoming more important than the journal is the number of structures that are available in the database long before publication: approximately 10% of structures have been marked as "to be published" for more than two years.[20]

The transition from writing a conventional scientific paper to submitting data to a database is that, as its name suggests, the database is concerned only with data. Reviewers do not scrutinize experimental procedures to see if they are satisfactory: database integrity may be ensured, as in the example of the PDB, by checks for internal consistency, but basically the database is concerned with the data rather than with methods of obtaining it. None of this affects the principle of reproducibility, because any conflicts that arise in sequences or structural parameters still must be resolved. The self-correcting mechanism of science remains intact.

CHAPTER

7

PUBLISH OR PERISH

I was an embarrassment to the department when they did research assessment exercises. A message would go round the department: "Please give a list of your recent publications." And I would send back a statement: "None."[1] Peter Higgs (Nobel Prize in Physics, 2013)

"If a tree falls in a forest, and no one is there to hear it, does it make a sound," is an irrelevant question by the standards of modern science. The laws of physics say that when the tree falls, it creates vibrations, and those vibrations propagate through the air irrespective of whether anyone is there to hear them.[2] But we might certainly ask whether a scientific paper has any meaningful existence if not even a single researcher ever cites it.

There are about 30,000 scientific journals.[3] They publish around two million articles each year. PubMed, a major database of the scientific literature, lists about 33 million scientific papers, mostly in biology and medicine. Each paper is the authors' means of validating themselves as legitimate members of the scientific community.

It would be unfair to say, "you are only as good as your last paper," but it is fair to say that scientists' reputations rest on the papers they have published. But here's the rub: is that measured by quantity or quality? The concept of the "least publishable unit" expresses a cynical view that the sheer number of publications is the most important criterion.

The number of papers published each year has exploded since the turn of the century. It would be wonderful if this represented a commensurate increase in scientific knowledge; but growth in the number

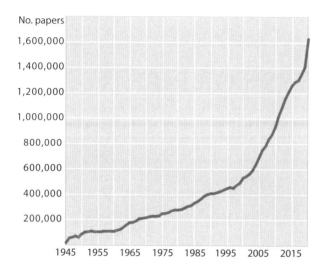

Science grew steadily through the twentieth century, but growth has been explosive in the twenty-first century.[4]

of papers outstrips the growth in funding for science, so it seems more likely to represent dissipation of knowledge rather than increased understanding.[5]

When I started *Cell* in 1974, the state of scientific publishing was primitive. Journals were more likely to treat submissions with benign neglect than to rush them into publication, unless special influence was brought to bear. When I was at *Nature* in 1971, the office of the Editor, John Maddox, was littered with submitted manuscripts that he had sequestered, and which the staff struggled frantically to retrieve. Far from enticing readers with an aperçu into exciting content, *Nature* still had advertisements on its front cover until 1974.

There was a giant divide between the general journals of science, *Nature* and *Science* and (perhaps) the *Proceedings of the National Academy of Sciences*, and more specialized journals devoted to specific fields. A paper could be published rapidly only in one of the general journals, and only in a rather truncated form paying no more than lip service to the notion that a paper should carry the information needed for its reproduction. *Nature* and *Science* did not even have a section for *Materials and Methods*. "Rapidly" is a relative term, meaning weeks or months, rather than years. Many of the most important papers in science really could not be fully assessed until much later, when more complete versions of the work appeared in specialist journals.

The cover of Nature, *Vol. 171, Issue 4356, for April 25, 1953 made no mention of the discovery of the double helix but had advertisements as usual.*

The idea behind *Cell* was to offer rapid publication (meaning a few weeks) for full-length papers, so that important observations could be seen to be true. It was axiomatic that we would handle submissions efficiently, rejecting as soon as possible if we did not want the paper, so it could be submitted elsewhere, and accepting (if necessary subject to revision) within no more than two or three weeks.

The scientific journal originated 350 years ago when the Royal Society of London started to publish the *Philosophical Transactions*. Early papers were mostly anecdotal; almost none would qualify as "scientific" by today's standards. The first example of anything that might qualify as a description of Materials and Methods was around 1744. By a century later, most papers were reporting experiments as opposed to anecdotes, with some account of how the work was performed. Although Robert Boyle was at pains to establish that the experiments with his air pump leading to Boyle's Law in the seventeenth century were reproducible, this was an exception: it was not

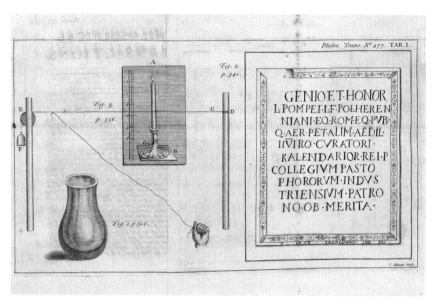

One of the first examples that might qualify as "Materials and Methods" was in a paper published in Philosophical Transactions, Vol. 488, *in 1744 on the fragility of glass vessels.*[6]

until the nineteenth century that the scientific paper really took its present form, with the focus on presenting data and detailing the methods of obtaining it.

Other societies followed the Royal Society, and today societies account for ~25% of scientific publishing. University presses entered the field relatively early and today are ~15% of scientific publishing. Commercial publishing was boosted in 1951, when Robert Maxwell bought the Butterworth-Springer house, known for "techniques of aggressive publishing in science"[7]—setting the model for decades to come—changed the name to Pergamon Press in 1951, and started a never-ending expansion. Elsevier, which had started its first international scientific journal, the infamous *Biochimica et Biophysica Acta (BBA)*, in 1947, bought Pergamon in 1991. (*BBA* was infamous for its huge volume, high price, and low quality.) Commercial publishers came to dominate scientific publishing. The share of the top five continued to increase, from 20% in 1973, to 50% in 2006, and 53% in 2013.[8] This was partly because they continued to start new journals and partly because they grew by taking over smaller publishers.

Scientific publishing was stuffy, but going along with the general attitude that science was an intellectual preserve, scientists felt this was appropriate. *Cell* broke the tradition by using a bright, modern, sans-serif typeface. We also put the Experimental Procedures at the end of the paper, instead of the middle, so the logical argument of the paper could flow without interruption. When the first issue arrived and I showed it to one of the scientists on the editorial board, he said doubtfully, "It looks very nice, but is it *scientific*?"

Scientific journals changed in response to changes in science itself, in particular because of developments that brought basic research closer to impacting directly on human medicine. Competition for "hot" papers drove publishing to become faster; attention was paid to what impact a paper had and to the overall impact of each journal.

Some publications are more equal than others. Scientific journals form a steep hierarchy. At the top are the "general interest" journals, whose ideal is that a paper should be of interest to all scientists (even though this is increasingly unrealistic as specialization increases). *Nature* and *Science* are the top general journals, accompanied by *Cell* for biology. *PNAS* (*Proceedings of the National Academy of Sciences of the United States*) is equally broad but not at the same level. The next group consists of journals devoted to more specific but relatively broad fields, such as cell biology or bacteriology.

Going farther down the hierarchy, journals become increasingly specialized and devoted to smaller and smaller subdisciplines, with some even focused on specific hypotheses or organisms. Picayune would scarcely be an adequate description. At the bottom of the hierarchy, you reach a point at which it's reasonable to wonder if the journal makes a useful contribution to science or merely decreases the signal:noise ratio. Papers usually start out by being submitted to a journal at some point in the hierarchy that the authors consider appropriate: rejection is likely to lead to submission to a journal lower in the hierarchy.

The first criterion in considering a submission is usually to assess its novelty. Conventional wisdom (or dogma in the sense of Chapter 15) has an especially important role here. By definition, a paper that conflicts with dogma should have general interest, but is also more likely to be rejected out of hand for being out of line with current opinion.

Usually you have to read most of a paper to form a judgment, but occasionally the introduction alone makes rejection inevitable. I think the most rapid rejection ever at *Cell* was of a paper in which the third paragraph concluded, "And thus we refute the second law of thermodynamics." (I have no idea why a paper on this subject should have been submitted to *Cell*.)

I rejected the paper without reading any further, but when I told my children about it that evening they were indignant that I had not considered the argument more fully. They were going through the American educational system at a time when relativism was all, and the thought of any absolutes was anathema. It was difficult to persuade them that the second law of thermodynamics is not dogma, but is immutable. There are times when the distinction between dogma and facts is not so clear.

It became conventional wisdom that the majority of published papers actually make zero contribution to science because they are never read by anyone except the authors, the reviewers, and (presumably) the journal editor. The figure that was floating around in 1990 was that 55% of publications are not cited in the five years following publication.[9] A more detailed examination confined to original articles (excluding other types of communication such as abstracts and editorials) showed that the proportion varied with the field:

- 22% in science,
- 48% in social science,
- 93% in the arts and humanities.[10]

Methodology is imprecise, and absolute numbers vary, but there is a consistent difference between medicine and science (with the fewest uncited papers), social sciences, and humanities (always at the bottom).

For articles on biology and medicine in PubMed, the largest database of research articles, 21% of those published between 2000 and 2015 had never been cited as of the end of 2021; another 2.4% were cited only by their own authors.[11] The PubMed database is extensive, so not surprisingly includes a fair number of somewhat irrelevant publications. The numbers are a bit better for journals published in the

United States or for journals listed in the MEDLINE database, which is more discriminating.

This is still a pretty fair indictment of the state of publishing in biology and medicine, especially when you consider that papers cited only once or twice are unlikely to have much significance. Only 50% of papers are cited five times or more. Is it unfair to conclude that around half of scientific research really makes little useful contribution and 10% makes no contribution at all?

Focusing more sharply on biomedical science, analysis of papers published from work that was funded by the U.S. National Institutes of Health (NIH) shows a much better picture: fewer than 5% of papers are uncited or cited only by their authors, and >84% are cited five or more times. Considering the intense competition for NIH funds described in Chapter 9, this may show that the granting system has reached the limits of the ability of peer review to predict success. In fact, a success rate approaching 95% seems a strong validation of the system (if you accept the significance of citation analysis).

Citations don't tell the whole story of significance, of course. The most highly cited papers tend to be methods; putting the median citation rate of 5 per paper into perspective, a paper from Ed Southern at the University of Edinburgh in 1975 reporting a standard method for transferring fragments of DNA has been cited 12,000 times.[12] Methods papers do not fit either the hypothesis-driven or data-driven model of science, they are perhaps more in the direction of technology, but as citations show, they have overwhelming importance.

The concept that citations measure the importance of an article was proposed by Eugene Garfield in 1955.[13] It was largely ignored, but in 1960 he put his theory into practice by founding the Institute for Scientific Information (ISI). In those days before the Internet, it was not so easy to keep abreast of the scientific literature, and the ISI published *Current Contents*, a weekly compendium of the content lists of scientific journals. People would look through *Current Contents* as their first resource for identifying articles of interest, and then go to find the articles in the journals. *Current Contents* made a profit, but the ISI's other activities, including the Science Citation Index (SCI), which consisted of volumes of citation data, ran up losses.

Garfield's famous invention was the impact factor, modestly introduced as, "the citation index ... may help a historian to measure the influence of an article—that is, its 'impact factor.'" The impact factor subsequently made its impact more for assessing journals than individual articles: basically, it gives the average number of times an article in the journal is cited in the period following publication.[14]

The top science journals (*Nature, Science,* and *Cell*) have impact factors in a range of 40–50; the top medical journals (*New England Journal of Medicine* and *Lancet*) are somewhat higher. Before the impact factor was invented, no one realized the full extent to which they dominated reports of the most important research results. There's no question about what role they play: there is a real question about to what extent other journals are signal or noise. Impact factors go all the way down to close to zero, at which level a journal is definitely no more than noise.

The group of the top three science journals is sometimes, semi-pejoratively, semi-affectionately, called the *CNS* (which in another context stands for central nervous system). The concentration of important research into so few journals is shown by the fact that many of the "hot" stories can be followed exclusively by reading the *CNS* journals. Three is

Citation analysis became so important that by 1979 the ISI was able to build its own headquarters in Philadelphia.

probably the minimum number to ensure that no one journal becomes too arrogant with its power. Of course, the edifice of science is scarcely built from three journals, but relatively few highly significant papers are relegated to more specialized journals.

The *CNS* journals are sometimes accused of increasing competition in science. This is like asking the cart to push the horse. It's true that they thrive on competition; it is doubtful they would be so dominant if competition was not so strong, but they reflect rather than cause it.

When an interesting paper was submitted to *Cell* in a field with which I was not closely familiar, I would read up the background from previous publications before deciding how to handle it. It was very rare that I needed to go beyond the journals of the "big three" or *PNAS*. They were the only journals we subscribed to. Indeed, only major institutions could afford to buy the more specialized journals, anyway.

PNAS, although it has published many important papers, has never rivaled the big three. It is more a contender for the best of the also-rans. *The Proceedings of the National Academy of Sciences of the United States* was established in 1914. The irony is that the National Academy includes the most distinguished scientists in America—membership is a highly selective process—but instead of supporting the peer review system, the Academy's own journal in effect held that members were above peer review.

Until recently, members of the National Academy could communicate their own work to *PNAS* with essentially no oversight. (Members submitted their papers along with two reviews that they had commissioned.) A member was limited to four "Contributed Papers" each year. Some of these papers were very good; some were not. In addition, a member could communicate two papers from nonmembers, accompanied by reviews. The process made for very variable quality.

PNAS began to modernize its procedures in 1995, when it introduced direct submissions: anyone could submit to the editorial office, and the paper would be peer-reviewed, with a decision made by an editorial board member (one of 200!). Members retained some privileges as they could still sponsor a direct submission from another author. Taking aim at the big three journals, an announcement in 2010 that these Communicated Papers were to be eliminated, said, "Academy members continue to make the final decision on all *PNAS* papers, unlike the process in place at such journals as *Nature, Cell*, and *Science*, in which final editorial decisions are often made by staff rather than practicing researchers." Here is a demonstration that the old attitude still exists: only practicing researchers should be involved in science.

Submission via members stopped in 2014. At this point, *PNAS* became essentially a peer-reviewed journal, functioning like the big three (with a 15%–17% acceptance rate for direct submissions), except for the feature that all decisions are reviewed by a member of the National Academy.[15] Members can still submit their own work as Contributed Papers.

About 25% of published papers are Contributed Papers. The numbers suggest almost all of them are accepted. They appear to be bimodal in their impact: the best are among the top 10% of cited *PNAS* papers; the worst are at the bottom.[16] Perhaps because *PNAS* gave them the notion they were above peer review, the most fuss about rejections at *Cell* always came from Academy members.

The absolute number of citations gives a slightly different view than impact factors (because impact factors do not take account of the total number of papers each journal publishes). Take the example of the new online journal, *eLife*. In its first decade of publication, 20% of citations in papers in *eLife* were to the *CNS* trio plus *PNAS* (which publishes 10 times more papers than any of the trio). Half of the citations were to fewer than 50 journals. The other half is made up of more than 8000 journals, of which 4000 altogether account for <1% of citations (each mentioned only once or twice). From the perspective of the thousands of scientists publishing in *eLife*, it would probably make no difference to science if that last 4000 journals ceased publication, and it would definitely make no difference if the 10,000 journals that aren't cited at all simply disappeared. In fact, eliminating those journals would benefit science by reducing clutter and noise.

Sociologists have debated whether major advances in science occur by the Newton effect or the Ortega effect. The Newton effect takes its name from Newton's metaphor in a letter to Robert Hooke in which he said that, "If I have seen further, it is by standing on the shoulders of giants." The Ortega effect is named for Spanish philosopher Ortega y

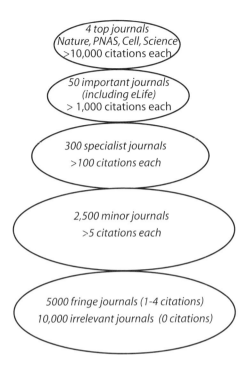

A view of the hierarchy of life science journals from papers in eLife. Four journals stand out as having the most citations, another 50 are highly significant, and 300 specialist journals are significant. Remaining citations are highly dispersed into journals of extreme specialization or little interest.[17]

Gasset, who argued that top-level research is like the tip of an iceberg, resting on a base of much less important research. "Experimental science has progressed thanks in great part to the work of men astoundingly mediocre, and even less than mediocre," Ortega said.[18]

Citation analysis suggests that highly cited papers tend to cite other highly cited papers.[19] This supports the Newton effect. At least as seen through the prism of highly cited papers, major advances come from an elite that to a large extent ignores much of the body of science.

Scientific publishing is concentrated into a small number of large publishers, but here the effect is almost exactly the opposite of the concentration of high-profile research into top journals. Larger publishers are famous for the lack of quality of their journals and their unscrupulous pricing policies. Known in the scientific community as the Evil Empire, Elsevier is probably the most hated publisher.

None of the top journals rose to prominence under the aegis of a major scientific publisher. Much like Big Pharma was unable to penetrate biotech on its own account, and entered the modern era only by buying up biotech firms, none of the big publishers appear to have been able to move into quality by themselves. Springer-Verlag bought *Nature* and used that as the basis to form a group of higher-quality journals. Elsevier followed the same strategy by buying the *Lancet* and Cell Press, including *Cell* and other journals.

Scientific publishing was a staid backwater until *Nature* shook things up in the early 1970s. The introduction of (initially anonymous) comments on papers published in *Nature* and elsewhere brought a sense of "news" to science itself as well as to the politics of science. It was a harbinger of things to come when *Nature* split into three journals in 1971. It probably seems completely unsurprising now, because many of the leading journals have effectively become brands, but it was revolutionary then.

The impetus was that *Nature* was getting overwhelmed with papers in molecular biology. So *Nature New Biology* was started to take the overflow, and *Nature Physical Science* was started to balance it. *Nature Physical Science* came out on Mondays, *Nature New Biology* on Wednesdays, and *Nature* itself on Fridays.

Things did not work out quite as planned. The original idea was that there would be only one route for submissions: they would all come to *Nature*, and we would decide which journal we wanted to publish them in. There was some concern that people who submitted to *Nature* would resist being relegated to a satellite journal. But fairly rapidly, people started submitting directly to *Nature New Biology*. This did not go down very well with John Maddox, the Editor of *Nature*.

The experiment with multiple journals at *Nature* was short-lived, and was wound up at the end of 1973. It took more than two decades for the concept to be resurrected. By the end of the century, *Nature* and *Cell* each had several associated journals, although the model was slightly different: *Nature* had a greater direct connection with its satellites; *Cell's* satellites functioned more independently.

One of the most significant changes in scientific publishing this century has been the transition from individual journals to groups. An obvious concern is that publishing satellite journals of more restricted interest, or lower quality, might lower the reputation of the lead journal, but in fact it has worked out the other way round (so far): the halo effect of the top journal increases the reputation of the satellites. In effect, *Nature* and *Cell* have become brands. The range of journals in each group is so broad as to extend well beyond simple satellites of the top journal: it's more a declaration that the publisher regards these journals as an exception from the low quality of most of their publications. Given the rising numbers in each group, however, I wonder whether the halo effect will be diluted.

Of course, societies often publish what amounts to a group of scholarly journals. The American Society for Microbiology, for example, publishes 13 journals. Society ownership is an imprimatur of a certain common quality level. However, the journals are edited independently (by researchers in each specialty), and there isn't any question of a halo effect from one prominent journal.

The largest five commercial scientific publishers own 15,000 journals, half of the estimated total of scientific journals.[20] Looking at the titles, you wonder what is the equivalent to the least publishable unit—the smallest definable scientific field, perhaps? The average impact factor for the journals of major publishers is between 3 and 5. University presses fare a little better; *PLoS* (an online publishing effort, discussed in the next chapter)

■ LEADING JOURNALS HAVE EXPANDED INCREASINGLY INTO GROUPS DURING THE TWENTY-FIRST CENTURY ■

	JOURNAL/GROUP	NUMBER	FORMAT
1974	*Nature*	Single journal	Print
	Science	Single journal	Print
	PNAS	Single journal	Print
1999	*Nature*	~6 journals	Print & online
	Cell	4 journals	Print & online
	Science	Single journal	Print & online
2021	Nature group	~168 journals	Print & online
	Cell group	~50 journals	Print & online
	Science	7 journals	Print & online
	PLoS	12 journals	Online
	eLife	Single journal	Online
	PNAS	2 journals	Online

achieves an intermediate level compared with the groups of journals spun off the high-impact journals, which vary from 9 for Nature Publishing (lower than the others because Springer is up to its old trick of over-expanding) and 17–20 for the Lancet group and Cell Press.[21]

An even more telling statistic is that the proportion of articles that are uncited or cited only by the authors is >30% for all the large commercial publishers (including Springer and Elsevier), compared with <20% for Nature Publishing, and <10% for *PLoS* and Cell Press.[22] The discrepancy is even wider looking at papers cited more than four times, which you might regard as the minimum level to indicate a significant contribution to science. The best you can say about the major publishers is that perhaps at most one-third of their papers have any tangible effect on science.

Springer now calls itself Springer Nature, perhaps to associate the quality of *Nature* with the poorer reputation of its other journals. (They even go so far as to say, "Springer is part of Springer Nature," an interesting reversal of reality.[23]) A revealing indication of its approach came from a mass retraction of 44 papers from the *Arabian Journal of Geosciences* in 2021. An example of a title of a retracted paper is "Distribution of earthquake activity in mountain area based on embedded system and physical fitness detection of basketball." This tells you all you need to know about the pernicious influence of some publishers on science.

The retraction (without any response from the authors) stated that, "The Editor-in-Chief and the Publisher have retracted this article because the content of this article is nonsensical." The "research integrity" director at Springer Nature said, "We are developing new AI and other tech-based tools and putting additional checks in place to identify and prevent attempts of deliberate manipulation." It is a sad view of scientific publishing to believe that artificial intelligence (AI) techniques are needed to detect rubbish. Whatever happened to the good old review process in which qualified researchers read the article.[24,25]

I propose two basic "laws" of scientific publishing: the damage that commercial publishers do to science is proportional to their size; and the damage that an individual journal does to science is proportional to its cost. The increasing number of journals, the narrowness of focus, and the lack of impact, all serve to decrease the signal:noise ratio in the scientific literature.

And there has for years been a spiral of decreasing circulation and increasing price. This started with a calculation: what price has to be charged to ensure that a profit is made if the journal is purchased only by the minimum number of institutions who feel they must have it? That price is inevitably high enough that some of those institutions decide they do not need the journal after all; so next year the price goes up to compensate for falling sales. If a paper is published in a journal that only a handful of institutions worldwide can afford, should it still count as a legitimate contribution to science?

I regard the large commercial publishers as a blight retarding rather than advancing science. Publishing huge numbers of journals with little regard for quality significantly lowers the signal:noise ratio. Charging exorbitant sums for access makes their contents available only to a privileged few at major institutions. And the costs waste institutional funds that could be better spent.

There is a counterargument to this litany of complaints that the unmanageable growth of the scientific literature favors quantity over quality. The signal-to-noise ratio in science is decidedly poor: the vast majority of scientific papers are not really very interesting. But you can't tell what apparently abstruse or arcane datum may spark a whole field of interest. To get the jewels, you have to support a lot of dross: the question is just how much is necessary?

The paper that Francisco Mojica published in 1995 on repeating units in bacterial DNA, which led to the discovery of CRISPR, was not cited by any other author until 2010: so any analysis of citation patterns for the first 15 years would have classified it as noise rather than signal. As I show in Chapter 18, CRISPR became one of the major technical developments of the twenty-first century. Of course, this is an extreme case, but it makes the point that we may need to tolerate a remarkably poor signal:noise ratio in science in order to be sure of catching all significant work. Whether it needs to be as poor as it actually is could be a matter for debate…

When I started in science, researchers were reluctant to talk to the press. They regarded the research paper as the principal, and sometimes only, means of communicating their work. As research came closer to impinging directly on human activities, especially when genes responsible for specific diseases were being identified, researchers were more likely to elbow one another out of the way in order to talk to the press after their paper was published.

But there was an accepted standard that you did not talk to the press until the paper was published, the principle going back to the basic view of science that all work must be reproducible by others in order to be accepted as part of the canon. It could hardly be considered reproducible, or even possible to assess, if the details had not even been published.

We applied the Ingelfinger rule religiously at *Cell*. The rule takes its name from Franz Ingelfinger, Editor of the *New England Journal of Medicine*, who, from 1969, refused to consider a paper for publication if its contents had previously been reported elsewhere. This applied specifically to releasing information to the press; it does not exclude presenting results at scientific meetings, which are regarded as informal presentations in a privileged context.

We took a pretty rigorous view at *Cell*, extending the rule to exclude any double publication. This meant we would not allow a paper that was published in *Cell* to be duplicated elsewhere, in particular in a book of symposium proceedings. At the time, these books were a blight on science: any number of publishers (dominated by the usual suspects) were looking to publish proceedings of symposia, and the books essentially contained nothing that had not already been published in the primary literature, but wasted resources if libraries were foolish enough to buy them.

In retrospect, it probably wasn't really our position to take a stand on moral grounds and perhaps we should have let the symposium proceedings die a natural death of their own causes. (We made an exception for the *Cold Spring Harbor Symposia*, which has always been a distinguished series, and in which there was something of a tradition of writing papers that brought together key original data from several research papers, thus providing a useful summary at the level of the data.)

The history of press releases of scientific information that had not been peer-reviewed and later proved to be wrong lends generous support to the notion that prior release of results is more likely to misinform people and cause damage than to fulfill an urgent public need to know.

The announcement of cold fusion, which offered a novel means of generating energy, merely misled people; but announcements of medical advances, in particular drugs that can treat disease but that ultimately turn out to be ineffective when properly tested, can do actual harm.

When science was a relatively arcane pursuit without much immediate application to human welfare, the policy was not controversial. Moving into the medical arena, the argument became a question of whether withholding information that might save lives was justified merely because a journal had not yet published the article. This argument had some force when it routinely took several months to publish an article.

Our answer at *Cell* was to publish important papers rapidly, in some cases as quickly as two weeks. Now, of course, papers can routinely be published online in a matter of days. Given sufficient speed in publication, it is surely better to have a more orderly process, when the full results are available for everyone to scrutinize, than for half-baked information to be presented to the public.

The situation became more acute with the AIDS epidemic, and the *New England Journal of Medicine*, for example, began to make some exceptions.[26] By the time of the COVID pandemic, the situation changed to the extent that publication via the *New York Times* or comparable organs seemed as common as publication in a peer-reviewed journal. We might in fact have been spared a significant number of false claims of possible treatments had the press declined to report on studies that had not been peer-reviewed, let alone even written up as a scientific paper.

Marcia Angell, a former Editor at the *New England Journal of Medicine*, regrets the change. "In the old days, there was much tighter control over a well-defined process. Optimally, research results went from the researcher to the journal to the media. Shortcuts were frowned upon, including press conferences by researchers or their institutions. Now everything can be out there immediately, with no filters. Some would make the argument that that is fine, because good work will rise to the top of the heap and it will all be faster. I'm not at all sure that's so. And the media don't want to wait. Now, for example, drug companies directly announce results to the media without waiting for peer review or publication. In short, I believe in filters."[27]

Publication by press conference became common when the fledgling biotech industry was trying to establish itself. Establishing priority became a matter of filing for patents and issuing a press release, or holding a press conference, rather than waiting for a journal to publish the paper. Often enough, the data weren't even complete enough to publish a peer reviewed paper. Here, as in other aspects of science, biotech led the way into the modern era.

All that said, critics can still maintain that journals' refusal to allow reporting prior to publication is more concerned with protecting their proprietary interests than serving the public good. Before the present era, I had an interesting exchange with Eugene Garfield, founder of the Institute for Scientific Information, about the extent to which entrenched interests might color one's views.

The ISI published an "Atlas of Science" in 1982.[28] It was a compendium of 102 "minireviews of research-front specialties," assembled on a somewhat unusual basis, described as an "innovation in information retrieval products." The driving force for selection was supposed to avoid the subjectivity of asking a reviewer to take on a topic; instead "bibliographies are identified objectively through the citation patterns of publishing scientists and not chosen by individual editors or contributors." The bibliography was assembled first, by identifying a group of papers that were both highly cited and co-cited (included together in reference lists). Then a scientist was found to review specifically those papers and only those papers.[29]

Reviewing the Atlas for *Cell,* I pointed out the main problem: the reviews were completely uncritical and took every assertion in every cited paper as gospel (even when some of the cited papers were devoted to refuting other cited papers). I guess my tone was a little hostile. "It is the very antithesis of what science should be about," I concluded.[30]

Garfield wrote an aggrieved response, well perhaps it was more rueful than aggrieved, arguing that my review was nothing more than an attempt to protect my position. As a journal Editor, with my livelihood depending on reviewing papers and commissioning reviews, of course I would react to any attempt to modernize the system as a threat. This seemed quite persuasive, and I began to wonder if indeed I was a Luddite who had been outmoded and encircled by superior forces.

Then I read the next paragraph. "Unfortunately," I remember it starting, "at the present we are forced to rely upon human reviewers to actually write the reviews of the cited papers, but in due course we should be able to replace them by software." I wondered why Gene had not gone on to the next logical step: to replace researchers with robots and algorithms. So I decided that the role of the Editor was not finished quite yet.

The moral is that we should recognize the limits of citation analysis. Cold fusion was one of the scientific fiascos of the century, but for that very reason, a highly cited paper. Citation analysis takes no account of *why* a paper is cited. It has some value in delineating the hierarchy of journals, but much less value in identifying the intellectual contribution of individual papers. Key papers in a chain leading to a Nobel Prize may be cited a few hundred to a couple of thousand times, but would be hard to pick out by citation analysis.

Nobel Prizes feature quite often in this book, not surprisingly because they acknowledge a good proportion of the most important discoveries. Almost, but not quite all, of the most significant discoveries discussed here were rewarded with Nobel Prizes. Criticism of the Nobel Prize most often comes from a feeling that significant contributors have been omitted. Nobel Prizes probably do more good than harm, but the good tends to be outside science, whereas the harm can be within it.

Publicity about the Nobels each year no doubt increases public awareness of science and sympathy for the scientific endeavor. Nobel Prize winners tend to become less productive after the award; this might be partly because they are no longer spurred by the ambition to gain a Nobel, and partly because they are distracted by other demands on their time. They can become elevated above their peers to a disproportionate level, and probably find it easier to get grants after the Prize (although their best work is behind them).

Nobel Prize winners beget Nobel Prize winners. Sociologist Harriet Zuckerman found that "more than half (48) of the 92 laureates who did their prize-winning research in the United States by 1972 had worked [in some capacity] under older Nobel Laureates." And 71 of the 92 prizewinners had students or postdocs who went on to become laureates themselves.[31] Zuckerman interprets this to show that Nobel Prize winners are a self-perpetuating elite, partly because very bright students pick Nobel Prize winners to work with, partly because the prizewinners can have their pick of applicants. Of course, there might also be another factor: Nobel Prize winners have a role in nominating candidates for future prizes and might favor their former students.

Nobel Prizes reward small-scale science. The limitation of a prize to no more than three prizewinners implies a view that significant discoveries can be made only by small numbers of people. The Peace Prize and Literature Prize are usually awarded to single individuals; the science prizes—Physics, Chemistry, and Physiology or Medicine—are more usually each split between two or three people.

The Nobel is by the far most famous of any of the prizes for science. Although there are now quite a lot of other prizes, they have a tendency to validate themselves by whether the winners later go on to be awarded Nobels. The Nobel Prizes are quite flexible with regard to subject matter: work in molecular biology has been variously rewarded by prizes for either Chemistry or for Physiology or Medicine. (Watson and Crick were nominated for both prizes: Physiology or Medicine might have been chosen because Max Perutz and John Kendrew, also from the MRC Lab of Molecular Biology, were awarded the Chemistry Prize that same year.)

It is felt fitting for prizewinners to express surprise, or at least humility, at the announcement they have been awarded a prize, but the fact is that the prizes are often subjects of active politicking. Some scientists campaign for years to get a Nobel. The stream of scientists giving seminars in Stockholm is out of all proportion to its position at the far end of the scientific food chain, but the Nobel gives it an influence far above its location or scientific contributions. Not the least effect of the Nobel Prize on science is the increased competition it engenders among scientists who feel they are in the running.

Going to a scientific conference in Stockholm can have aspects of the surreal. Local researchers come up to you and whisper, "I have influence with the Nobel committee." Participants at the meeting from elsewhere sidle up to you and say, "I have just been told that I am in line for a Prize." You wonder how they can be under an illusion with so little connection to reality.

Nobel Prizes for science are a real mix. About half go to researchers who everyone would agree should have a Nobel. About a quarter are debatable, at least in the sense that there is other equally distinguished work that would have been just as deserving. A few awards are quite frankly mistakes: not usually bad science, but simply not as significant as the committee thought.

Science does not necessarily progress smoothly. I know of suspicions concerning Nobel Prizes awarded while I was Editor of *Cell*: in one case, it is probably impossible that the results in the key paper could have been obtained as described, and, in another, the interpretation may have been based on learning of the results of a competitor. Yet both prizewinners are deservedly among the most respected scientists of their generation for publishing bodies of work of undoubted significance.

The scientific literature became much livelier in the last years of the twentieth century, with many more journals adding reviews and commentaries to research papers. Technological improvements made it possible to publish papers in weeks rather than months (or even years). But the basic model stayed the same: the research paper was the unit of publication, and the process of getting from manuscript to published article remained very much the same. Things began to change after 2000, when the Internet became more than an alternative medium for the same message and began to change the way that science is communicated, as I discuss in the next chapter.

One thing that has not changed, and that could not change without a corresponding change in the very fabric of science, is the need to publish your results. Sometimes publication becomes an end rather than a means. Occasionally—very occasionally, actually—we would receive a submission at *Cell* in which the authors had made an interesting observation, but misunderstood it to the point of getting their conclusions the wrong way round. We might accept the paper but require the Discussion to change its conclusions; in fact, on one occasion I even wrote a new Discussion for the authors to reverse the conclusions. It surprised me when the authors accepted the suggestion without any argument, when I had expected them to fight for their original conclusions. It was hard to decide whether they were exhibiting the true scientific spirit—accepting where the data led—or abandoning their principles. It worried me that they seemed to care more about getting published than their conclusions.

The origins of the phrase "publish or perish" are not clear, but it may date from 1942.[32] It represents a truism that a scientific career can be summarized by the number of published papers. Peter Higgs, a theoretical physicist who predicted the existence in 1964 of what became known as the Higgs boson, published fewer than 10 papers subsequently. Theoreticians can get away with fewer publications than experimentalists, but Higgs said, when he was awarded the Nobel Prize in 2013, that Edinburgh Universality had considered firing him, and that "Today I wouldn't get an academic job. It's as simple as that. I don't think I would be regarded as productive enough."[33]

It's difficult to measure quality and much easier to measure quantity. The two get confused in the idea that impact can be measured through citation analysis, and the use of impact factors (or other comparable measures) to assess candidates for tenure is responsible for much of the hold of citation analysis on research. There is huge pressure when you are a candidate for tenure to have a paper in one of the big three *CNS* journals, and significant bonus points if the paper is featured on the cover.

Experimentalists are expected to keep up a steady flow of experiments. The pressure to publish comes from the institutional structure of science or, to be more exact, from the view of the organizations providing funding that they must show some return for their expenditure. Combined with the reluctance to regard negative results as significant achievements, this in turn directs research toward safer topics, designed to produce a predictable flow of research articles in the short term, as opposed to riskier leaps into the unknown that might or might not produce results in the long term.

It would be unfair to say this attitude reflects a change in science, because it's been the dominant influence ever since funding became institutionalized, although it's enhanced by increased difficulties with funding, as discussed in Chapter 9. It suppresses the sense of individuality, and means "bigger" research is safer than smaller-scale research. Even if science does have a unique value system that continues to define it as an intellectual activity, the enterprise as a whole cannot escape the influence of the institutions in which it is organized.

CHAPTER

8

E-SCIENCE

Privately owned journals can't tie up scientific information.[1]

Harold Varmus, 2003

It is not written in the stars that we have to have science publishers, and it is not a moral imperative for us to defend them.[2]

Vitek Tracz, 2005

I f the transition to electronic publication proves anything, it is that the medium is definitely not the message. In fact, the message has remained remarkably unchanged—extended and expanded perhaps— but basically similar in principle to that of a century ago. Electronic publication should have been a far more disruptive technology for science than it has actually proved to be. Does this tell us something about the practice of science, or is it more a comment on the sociology of scientists?

The research article is the lifeblood of science. Research acquires meaning when it is published for everyone to see. But the research article is more, much more, than the intellectual unit of science. Scientists are judged on their record of publications: reputation, funding, and tenure all depend on it. The research article has become the unit of assessment. So the form that the article takes is no small thing. In its original form, it was very much restricted by the means of distribution via a printed journal that comprises a collection of articles. Any substantial change in that mechanism challenges the status quo, not merely for the publishers and scientific societies who run the system, but for the very fabric of science.

There is almost no argument except nostalgia for publishing scientific articles on paper.[3] But for a system that by definition is at the front of modernity, science has not been so quick to change its mode of communication. Commitment to tradition is shown by the slow move from adding an online website to a print journal, to regarding the website as the primary means of publication. I'd say at a guess that this took about a decade longer than one might have expected. When will journals completely abandon print in favor of electronic format?[4]

In all fairness, it was not at all clear when online publication became possible that this would be the preferred mode of the future. We started online at *Cell* by making abstracts freely available in the mid-1990s. When we moved to a full online site, it was not obvious whether it should be included with a print subscription or should be an additional charge, or whether it should give access only to the current year or also to archival issues. The model may seem obvious now, but at the time all this was open to question.

Online publication offers a dramatic change in distribution: it is no longer necessary to physically print and distribute the journal. Anyone with a connection to the Internet can have immediate access. This poses an enormous threat to the traditional business model, in which access was essentially restricted by the small number of copies—often a single copy in a library was the only source. The solution the large publishers came up with, charging huge fees for a subscription or for single access to an article, spawned a rebellion in the scientific community.

It was nothing less than a call for revolution in 1999 when Harold Varmus, Director of the NIH, proposed the establishment of a central site for electronic publication, which he called e-BioMed.[5] The response was appropriate to such a clarion call, ranging from vitriolic to messianic.

e-BioMed would be established by the NIH to offer two modes of publication: a traditional peer-review mechanism overseen by editorial boards; and a general repository bypassing peer review. All reports on e-BioMed would be available free online. Authors would own the copyright of their own material. It would be an addition to existing journals, although archival material from existing journals could be added to e-BioMed.

The proposal for a general repository was widely trashed for the idea that papers could be presented without peer review, although physicists have been using the arXiv system to circulate preprints (papers that have not yet been submitted to a journal) since 1991. In fact, NIH had tried an experiment along these lines from 1961 to 1967, called Information Exchange Groups (IEGs). There were several IEGs for relatively specialized topics, and the system functioned by copying and mailing the articles to participants. *Nature* and *Science* objected to the system, in terms that were condescending if not vicious, but the *coup de grâce* came from a meeting of editors of biochemical journals, mostly owned by commercial publishers, who decided they would not consider submissions of papers that had been circulated in IEGs.[6]

Half a century later, preprints in biology are being circulated by the bioRxiv server run by Cold Spring Harbor Laboratory. (There's a similar preprint server for medicine, medRxiv.) To put this into perspective, bioRxiv has circulated about 40,000 preprints per year, more than an order of magnitude less than the number of published research articles, but the trend is clear that it may become the rule rather than the exception for work to be circulated as a preprint before it is formally published.

The status of the preprint is a curious halfway house: it's not a privileged communication (the data are out there and anyone can make use of them), but it doesn't carry the imprimatur of a reviewed article (although that is admittedly variable). It's probably preferable to cite a preprint, allowing a reader to form their own impression of its quality, than to use the old format of "unpublished data" or "personal communication," which leaves the status of the information in obscurity. The problem is more with reporting to a general audience, when the distinction that a preprint has not been reviewed is easily lost.

The proposal for a peer-reviewed mechanism via NIH attracted support from scientists (in contrast to the adverse reaction toward a general repository) but hostility from vested interests, both commercial publishers (of course), but also nonprofit publishers such as societies (for whom profits from journals fund an important part of their activities). Commercial publishers sent their lobbyist to tell Congress that Varmus was "trying to undermine the free enterprise system by turning the NIH into a federal publishing company."[7]

This put the kibosh on the proposal for the moment, but later in 1999, PubMed was extended into PubMed Central, a digital library

■ TIMELINE OF SCIENTIFIC PUBLISHING ■

YEAR	MILESTONE
1665	Royal Society publishes *Philosophical Transactions*
1869	Macmillan publishes *Nature*
1880	AAAS publishes *Science*
1892	Chicago University Press publishes first journal
1906	Oxford University Press publishes its first medical journal
1915	National Academy publishes *PNAS*
1947	Elsevier publishes *BBA*
1997	PubMed released
2003	*PLoS* launches open-access online journals

based on articles provided by publishers of conventional journals with a delay of a few months or a year after publication. It attracted few contributions from publishers at first. Varmus regarded this as "unacceptable obstructionism," given that the delay meant there was unlikely to be any impact on revenues for the publisher anyway. But perhaps they could see what was coming.

Twenty years on, PubMed Central has links to free full text for a substantial proportion of papers, but the large commercial publishers still try to get away with the minimum possible. Researching for this book, I needed to consult a large number of older research articles. I encountered a barrage of paywalls on the publishers' websites. Except for the oldest, most could actually be obtained from the authors, but I suppose that for the publishers, it's a principle. This reinforced my view that, far from having any role in promoting science, the large commercial publishers are the biggest single impediment to its progress.

Try to access an article from *Nature* (Springer) or *Cell* (Elsevier) from the 1970s, for example, and more likely than not you will hit a paywall. This is beyond pointless: fifty-year-old articles are of archival interest, the potential income from selling them must be miniscule, but the scorpion cannot change its nature. (Not that this shortsightedness is limited to commercial publishers: the *New England Journal of Medicine* allows very little free access—the journal is thought to provide most of the $100 million its owner, the Massachusetts Medical Society, makes from publishing.[8])

The *New England Journal of Medicine* was firmly against e-BioMed. "The proposal fails to recognize fundamental differences between basic science and clinical journals," was Arnold Relman's opening criticism in his editorial.[9] He argued that science journals are read by researchers who understand the caveats, whereas medical journals are read by clinicians who "know little or nothing about the methods of published studies, and they depend on the accompanying editorials in clinical journals to help them interpret the data."

This is not a comforting view of the competence of practicing physicians, to say the least. Another reason emerged later in the editorial, that an "e-BioMed system that included clinical studies would very likely reduce the submissions, paid circulation, and income of most clinical journals enough to threaten their survival."

Funded by the NIH, e-BioMed was to provide open-access publishing, with no charges to users. Inspired by the concept of open access, publishing entrepreneur Vitek Tracz in London started BioMed Central in 2000. Of course, if there are no subscription fees, someone has to pay; in this case, the income came from fees to authors or advertising.

BioMed Central included general journals of biology and medicine as well as more specialized journals. It grew to 70 journals, all functioning by peer review, with the unusual feature that signed reviews were posted alongside the articles. This was an interesting experiment, but lost much of its purpose when it was sold to Springer in 2008.

Real open access (meaning that it did not need to charge fees in order to make a profit) started with the formation of *PLoS* (Public Library of Science) in 2003. The prequel was an open letter (subsequently signed by 34,000 scientists) in which Harold Varmus, Pat Brown (Stanford University), and Michael Eisen (University of California, Berkeley), urged scientists to publish or review articles only for journals that made their content free within six months of publication.[10]

When this had little effect on freedom of access, they obtained funds from a charitable foundation to start a publishing effort. Its flagship journals are *PLoS Biology* and *PLoS Medicine*, supported by another nine more specialized journals (all in biology or medicine), and an entirely general journal called *PLoS ONE*. Submissions are peer reviewed in the usual way. (Because profit was not the objective, there was no need to expand into every crevice of scientific publishing as a large commercial publisher would do.) The initial grant covered startup costs: authors are charged fees to cover the cost of publishing each article.

PLoS Biology and *Medicine* have been quite successful, even if they have not achieved the objective of competing with the top journals. Impact factors are 8–11 (a good specialist journal gets ~5), compared with 40–50 for *Nature, Science,* and *Cell.* The founders would probably regard this as an inappropriate criterion. "The excessive attention given to this pseudoscience [citation analysis], especially in Europe and other parts of the world, is having a noxious effect," says Harold Varmus.[11]

Varmus goes further when asked if he is satisfied with the impact of *PLoS* on scientific publication. "We haven't made as much progress as I had hoped towards dethroning the big three—*Science, Cell,* and *Nature.* Good as they are as journals, many of my colleagues use the appearance of articles in those three journals and their subsidiaries as an overly simplified certification of quality, creating an imperative to publish in them, distorting many aspects of academic careers in science."[12]

Perceptions that the large commercial publishers practice predatory pricing have led to various calls for boycotts. Peter Walter and Keith Yamamoto at UCSF (University of California, San Francisco) issued a call to boycott all Cell Press journals in 2003 because of the exorbitant fees Elsevier demanded to make electronic versions available to the university. "We can all think of better ways to spend our time than providing free services to support a publisher that values profit above its academic mission," they said.[13] (It was perhaps too kind to accept that the commercial publishers actually have any academic mission.) There has in fact been a call for a boycott every few years, but the top journals are so entrenched that these calls rarely have much effect.

You would think the move to electronic publishing would lead to a more measured assessment of the value of individual journals. When a physical copy of a journal is available in a library, it's not obvious how often it is used. Some journal subscriptions may have been purchased because they were of interest to a faculty member long ago, and there may be inertia in continuing to subscribe. But when articles are accessed via a library website, there is an immediate measure of how often a journal is accessed (or whether it is ever accessed at all). This is probably the greatest threat to the large publishers from the move to an electronic era.

So how can publishers stop libraries from simply canceling those journals that are rarely used? The answer is the bundle: all the publisher's journals are available through a single portal, and they are not sold separately. No cherry-picking here! "Never mind the quality, feel the width" is the model. But any model based on subscriptions is now threatened.

It took several years after *PLoS* started open-access publication for the model to spread to other journals. The biggest impact on open access came from a top-down regulation: from 2008, any work relying on an NIH grant has to appear in an open-access source within six months of publication. The HHMI in the United States and the Wellcome Trust in Britain, both major funders of science, had adopted the same policy in 2007. This means that most of the most significant research published since then in the United States and the United Kingdom is now available as open access. As most of this work has been published in subscription-based journals, compromises have been developed to allow open access while maintaining subscription income.

The initial compromise at most publishers was to delay open access as long as possible—that is, for six months; although this might eliminate income from sales of archival articles, it should have little effect on current subscriptions. It was really far from satisfactory, because it was not universal. Even now, for older papers, it's impossible to tell before you try to retrieve an article whether access is available. What is really needed is for every article on PubMed to have a link to the (free) full text of the article. This is (finally) happening for current articles, but not necessarily for archives.

Progress toward immediate open access comes from the *force majeure* of Plan S, a proposal that originated with major European funding agencies in 2018 and attracted other supporters. (S stands for "shock.") Any research funded by a participating agency must be freely available online immediately on publication. (This is sometimes called green open access.) If the journal does not allow this itself, authors can post the article in another online repository. Plan S also requires that the work can be copied and translated. This has been made a condition of grants as of 2021. It's been reinforced by a requirement in an OSTP (White House Office of Science and Technology Policy) Memorandum of 2022 that all federally funded research in the United States must be made freely accessible no later than December 2025.[14]

The move to open access has spawned a revolution in the economics of scientific publishing. Green open access effectively means there is no need to subscribe to the journal, so the principle that readers pay for the journal has been abolished. The inevitable consequence is that authors must pay.

There has always been something of a break between journals that have editorial staffs who directly take decisions on papers as opposed to journals for which decisions are taken by an external editor (typically a researcher in the specialty). The trade-off here is that the increased salary costs of an editorial staff have been compensated by the larger circulation (meaning greater subscription income) for journals with broader appeal. But that becomes problematic if publication becomes open access, so that subscription income is eliminated or reduced.

This entails something of a reversal of the traditional cost relationships. It used to be the case that a general-interest journal would have two major cost centers: salaries and production/distribution. A restricted-interest journal would have editorial costs that were insignificant compared with production costs, which would be in proportion to its size and circulation. So the change to electronic publication has a greater benefit for the restricted-interest journals because they have all but eliminated distribution costs, and much reduced production costs, whereas general-interest journals still have to pay for all those editors. This should have enabled restricted-interest journals to offer significant price reductions: of course, that has not happened.

Individual subscriptions are a significant proportion of subscriptions for general-interest journals (especially the *CNS* trio), but most restricted-interest journals are purchased solely by institutions. This had the effect of protecting individuals from the high subscription prices of commercial journals. But with an open-access model, every author has to ask whether the price is worth paying.

At one time, some journals (mostly from nonprofit publishers) required a page charge upon publication—typically anything up to a few hundred dollars per page published. This was a terrible principle: it always seemed to me that if you couldn't manage to publish a journal that attracted enough subscribers to keep its price reasonable, it probably wasn't worth publishing. (I admit that at one time, the rapidly increasing size of *Cell* forced imposition of a page charge, but this was an emergency situation, and the period was kept as short as possible.)

The modern equivalent to the page charge is the APC or Article Publishing Charge. This varies, depending on whether it has to compensate for potentially complete loss of subscription income from green open access, or partial loss because open access is delayed. Typically it runs anywhere from $2500 to $10,000 per article, depending on the journal and the publisher.

Some journals have released information about their costs to assuage authors' indignation. The *EMBO Journal* (one of four journals published by the *European Molecular Biology Organization*) is an example.[15] Current revenues are roughly two-thirds from subscriptions and one-third from APC. Current costs are roughly 40% production and 60% editorial. Publishing turns a profit that's just a bit less than the income from APC; this goes to support EMBO activities. So at present, subscription revenue almost covers costs.

Forced by Plan S, commercial publishers have conceded that they will offer authors the option that a paper will be made freely available for a fee. So the mixed model, with revenues partly from subscriptions and partly from APC, has become common. The problem is that as the balance moves more toward APC, the charges on authors from commercial publishers seem an increasingly less reasonable expense for the public purse, as it becomes clear that the APC is basically determined by rapacious profit margins. Perhaps funding agencies should impose a limit on payments to commercial publishers.

The hybrid model (mixing subscriptions and APC) is likely to be short-lived, because a revision to Plan S in 2019 refused to allow funding for journals with a hybrid model unless there is commitment to make a transition to full and immediate open access. A transitional period runs until 2024. (A derogation allows publication in a subscription-based journal so long as the work is also made available by open access elsewhere—for example, on a website at the authors' institution—but that rather undercuts the need to subscribe.) It is not obvious how commercial publishers will be able to make a successful transition to supporting their revenue stream principally by charging authors.

The barrier to entry for starting a print journal is significant: physical copies have to be printed and distributed, with heavy costs incurred long before any reasonable number of institutions subscribes. I remember that when the first issue of *Cell* was published, there were barely more than 100 subscribers.

Barrier to entry in electronic format is much lower: direct costs are much less and no distribution is required. So it would be reasonable to expect anyone who wants to start a latter-day *Cell* to do so in electronic format. One of the first signs that people might be thinking along these lines after the establishment of *PLoS* came with the formation of *eLife* in 2012 as a peer-reviewed, open-access journal. Funds were provided by the Howard Hughes Medical Institute, Max Planck Society, and the Wellcome Trust.

For all that barriers to entry are nominally much less than for print, it took $9 million to start up *PLoS* and $26 million to start *eLife* (for the period until 2016). The backers provided another $25 million for the period from 2017 to 2022. I don't remember exactly what *Cell* cost to start, but I think it was more of the order of $100,000—all done more or less on a shoestring. Shoestrings seem to have gone out of fashion.

The original idea for *eLife* was to make a high-impact journal as open access with no fees to either authors or subscribers. However, an APC of $2500 was introduced in 2016. Editor Randy Schekman criticized *Nature* and *Science* as "luxury journals," and wrote an article in *The Guardian* newspaper under the headline, "How journals like *Nature*, *Cell*, and *Science* are damaging science," which attracted some criticism, as these were the journals in which he had published his Nobel Prize–winning work.[16] For all that criticism, *eLife* has an acceptance rate of ~15%—only a bit higher than *Nature* and *Science's* stated 10%.[17]

eLife refuses to allow impact factors to affect its function. "We have opposed the use of the impact factor since day one because we feel it is meaningless, particularly when it is used to assess individual papers or scientists," Schekman said.[18] That said, it does not come out badly, somewhere around the level of *PLoS*. Mike Eisen, who was of the founders of *PLoS*, took over as chief editor in 2019.

The approach is nothing if not innovative. Standing on its head the original opposition of established journals to preprints, *eLife* announced that from 2021 it would consider *only* submissions that had previously circulated as preprints.[19] (At the time this was announced, 70% of papers under review at *eLife* were already available on one of the preprint servers.) This follows the concept of big data in assuming that a sufficient mass of criticism can substitute for the directed efforts of two or three reviewers: of course, it presupposes that there will be enough criticism, and that the authors will respond appropriately (not necessarily the case following peer review for journals).

The objective is to "break free of the 'one paper, one journal, one publication model'," by opening up the process. Any way you look at it, this is a far more transparent model for the process of submission and publication.

Although it is a strength that transparency is increased by requiring exposure as a preprint as a condition for publication, it is a weakness that the same exposure can lead to media reports of unverified data. Whereas researchers in the specialty for whom a preprint is intended may be able to assess it, the media is more likely to take it at face value. Of course, peer review is a variable process, and there is a spectrum from the rigor of a top journal of the field to a journal in which the review process is not much more than nominal. The point here is that the circulation of preprints extends that range in the direction of a lower signal:noise ratio.

One side of open access is exactly what the name indicates: all published work is accessible. The other side is that reduced barriers to entry make it easier to publish garbage. And just as science itself can slip from being careless to being fraudulent, so journals can pass from offering a poor price:value ratio or signal:noise ratio into being predatory. A bit like pornography—you know it when you see it, but it's hard to define— predatory journals are a recognized, but ill-defined phenomenon. Attempts to define them tend to include a list of warning signs—lack of transparency about editorial and pricing practices, excessive attempts to solicit submissions, and so on—but end up with definitions that are too broad, if not fatuous, to be useful.[20] You might view predatory journals as an extreme example of a vanity press, in which articles are published in exchange for payment; it goes without saying that there is no meaningful peer review.

Lists of predatory journals tend to be short-lived, because of legal actions and other complications, but one recent compilation had more than 1100 publishers publishing more than 1500 journals.[21] The general mark of the predatory journal is that it takes advantage of the open-access model, making its contents available to readers via the Internet without charge, but profiting from charges (often unexpected) to contributors. Like spam in e-mail, predatory journals are hard to eradicate.[22] They are an unfortunate side effect of the otherwise beneficial move to make the scientific literature available to everyone.

Electronic publication started as an add-on to print publication, by making the paper that had been published in print available online. The online version became more valuable when it was realized that it could be free of the constraints that limit print—for example, papers could have links to Supplemental Material. Online journals don't have volumes, don't publish issues, and there are no contents lists: they just publish papers. It does mean you have to work harder to find the papers that interest you.

Vastly increased speed of publication has been one major effect of online publishing. Until the 1970s, the delay from submission to publication was usually measured in months, if not years. One objective of *Cell* was to move that timeline to weeks. Since the turn of the millennium, the definitive moment has been the publication online, at first some weeks ahead of print (essentially limited only by the length of the review process)—today sometimes the sole form of publication.

The latest development is the possibility of accessing submitted papers while they are in the review process. Cell Press has a feature called *Sneak Peek* and Springer Nature has a feature called *In Review*, both introduced in 2018. If the authors agree, a submitted paper appears on the server within a few days of submission, much as it might appear on a preprint server. The pros and cons are much the same—the advantage is that your claim to novelty is immediate; the disadvantages are that you have alerted rivals, and there is risk of embarrassment if you have made a mistake—but it is also true that, if the paper is rejected, everyone knows that the journal has found fault with it.

Appearance in either format undercuts even more substantially than a preprint server the concept that submission to a journal is a privileged communication, known only to the editors and reviewers. On the one hand, this opens up a real can of worms for claims regarding priority: what is your position regarding priority if you submit a paper, make public your claim, a reviewer finds a fatal flaw so the paper is rejected, you fix the flaw, resubmit, and are published; but in the meantime, another laboratory takes advantage of your (flawed) report to complete and publish its own report earlier. On the other hand, the increase in transparency reduces possible abuses of the process.

Cell Press's Sneak Peek uses Elsevier's SSRN server, basically a giant preprint server, which provides a mix of free and paid services. Separating articles into individual "journals" is somewhat nominal, as basically this is a giant melting pot. This is the logical evolution of the approach of the giant commercial publishers: publish anything, and make money at it. I wish I could quantify how much it has reduced the signal:noise ratio in science. It is Harold Varmus's dream of free communication among scientists turned into a nightmare.

There is still a line between published (reviewed) papers and preprints. But if the line becomes blurred further—especially if the media do not draw a distinction—might scientific publishing descend into a free-for-all? This brings us to the question of the roles of research articles and individual journals.

With the Internet why do we still need publishers to distribute journals? In fact, why do we need journals at all? Aside from their role in distribution, journals have functioned as gatekeepers that offer an imprimatur of quality (the quality varying with the reputation of the journal). Journals in specific fields provide a means of identifying advances in those fields; whereas for general journals, the claim is that they identify papers that are of such interest they should be read by researchers in other fields.

The reputation of the gatekeeper may either boost or reduce the impact of a paper. The scientific literature is now so vast that searching for any topic produces far too many results. Starting with a journal of known reputation narrows the results to provide a reasonable starting point. Should the role of gatekeeper now be unlinked from the role of the journal as the means of distribution?

In the technical sense, if papers are to be presented on a website, the journal becomes irrelevant as a distributor. A publisher could offer sophisticated means of identifying papers in specific areas, which for researchers trying to keep up with specific fields would be more useful than the divisions of conventional journals. Indeed, a halfway house between the journal and free-for-all is that some publishers offer "collections" of papers online, presenting all the papers from all their journals on a given topic.

With the general expansion of the literature, and the move to make everything accessible immediately online, a gatekeeper function has the attraction of holding back the deluge. Remember the statistic that many papers are never, or rarely, cited: that probably means that they have never even been read. Just having to go through them all in order to dismiss them would be a significant burden. The gatekeeper avoids this by relegating them to a journal most people never read. But if we set out to design a gatekeeper function for the modern era, would it take the form of a journal?

It is not yet clear what type of gatekeeper function might replace the journal. Perhaps some Google-like mechanism, offering an assessment in terms of reader reactions? Could the circulation of preprints, perhaps combined with some sort of open-review process that adds the reputation of the reviewers to the equation, lead to a software-driven assessment of a paper? After all, the reputation of a journal depends in no small part on the rigor of its review process and the selective criteria this applies: does this necessarily need to be associated with a journal in the sense of a discrete publisher?

If journals disappear as a means of collecting papers of similar subject and quality, some equivalent to *Current Contents*, the means of identifying papers of interest in print journals that preceded the internet, as discussed in the previous chapter, might be resurrected. Assessment of quality and interest might become divorced from the means of publication.

Online publishing has made great advances—it has changed the whole gestalt by making open access possible. But it has otherwise so far been much less disruptive than you might have expected from the possibilities of this new technology. That has not generally been due to any overt resistance to change from scientists; the stumbling block has been more the success of the large commercial publishers in protecting their interests, basically in making the minimal modifications to the publishing model required to head off the threat to their profit line.

Why has e-publication disrupted publication (by changing pricing models via open access), but failed to have much impact on the format in which results are presented? Surely the first thoughts about going online would be that restrictions on space would become irrelevant, there would no longer be any limitations on size or color of figures, material could be updated if new information became available: and for that matter, comments or reviews from others could be added. In short, the stylized format of the research article would be regarded as *so* twentieth century (if not nineteenth century). The most significant tangible effect, however, has been the ability to include supplemental information, instead of citing it cryptically as unpublished data.

For the most part, online publishing has mimicked the format of the individual article, which really was dictated by the needs of print. Communication in general has been changed enormously by the growth of electronic formats, but science is still stuck with the concept of the individual research article. It's more flexible online than in print, but it's still an artifice to divide a continuous body of work into discrete papers. Then the question becomes whether a contribution should be regarded as a snapshot in time or should be updated as further information became available. It's still too early to see how this will fall out.

What would scientific communication look like if it were designed from scratch for the electronic era? Perhaps the single article should be replaced with an (online) thread (like Twitter)? You could start a thread when you become an independent investigator. Additions to the thread could be made when there was a sufficient advance and would establish priority in the same way as today's research article. Maintaining a thread would make it easier to follow the work and would avoid wasteful duplication between successive papers. A major discovery would be needed to start a new thread. Threads would probably most often branch off existing threads. (I recognize that I am ignoring here the aspect of big science that new contributions may be made to databases rather than as individual papers.)

A system of threads would have an enormous advantage that the context of an addition to the thread would be immediately apparent from the preceding thread. I cannot possibly count how many times I have read the introduction to a research article that started with something like "it has been established that...," only to find when I checked the relevant references that what had actually been established was really a bit different. This sort of spin would be easily eradicated when new information was presented as part of an existing thread or a branch from one.

Peer review could still be involved, although a new mechanism might need to be devised for consulting reviewers. There would need to be some form of substitution for the reputation of the journal in order to assess the rigor of the review process for each addition. Your reputation would depend on how many threads you were able to start. That's really not so different from the way things work at the moment, but would bring the process of publication into the modern era. What role there would be for publishers or journals in the current sense is an open question, especially because a thread would not necessarily be confined to a single journal or its equivalent. Perhaps the journal would become more of a verifying stamp of approval. It would be a brave new world if we could replace the centuries-old view of the scientific paper as the key goal of science.

SOCIETY

CHAPTER

9

FUNDING

A new scientific truth does not triumph by convincing its opponents and making them see the light, but rather because its opponents eventually die, and a new generation grows up that is familiar with it.[1]

Max Planck, 1949

Funding of science in the United States is the envy of the world. The NIH (the National Institutes of Health) spends billions of dollars a year on basic research, philanthropic organizations and universities spend significant amounts, and this is dwarfed by spending on applied research and development by industry. Yet scientists feel themselves beleaguered by the difficulties of obtaining grants. If there is any common perception, it is that the system is broken and no fix is in sight.

Let's backtrack through a research scientist's career. The process has scarcely changed in the past 100 years. Starting as a PhD candidate, you learn how to conduct a research project. Then you move on to a postdoctoral fellowship. With the increasing move to team rather than individual efforts, the principle that you function independently may be honored more in the breach than the observance. You may be tightly bound into the work of a group or allowed some degree of independence, depending on the laboratory. At the end of this period, you apply for a faculty position and set up your own laboratory, usually taking some project with you.

Once you have your own laboratory, you need to provide the driving force for directing a small team, with the objective that over the next five or six years you will produce a sufficient body of work to persuade the institution to grant you tenure. At that point, you have security, but you still need to keep the momentum going by publishing papers and attracting funding.

127

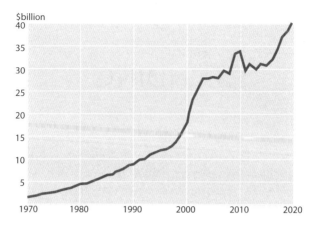

U.S. government funding for research in life sciences increased sharply because of a five-year plan to double spending on NIH between 1998 and 2003. Most spending goes to biology ($19 billion today), followed by medicine ($14 billion), with smaller amounts to agriculture or environment. Life sciences received only 28% of total government spending on research in 1970 but received 48% in 2020. This includes basic and applied research and development. About 10% of NIH funds are spent on the intramural program (funding research within the institutes at NIH) and 90% on funding grants at other institutions.[2] Of course, not all the billions given to extramural research are spent directly on science: about one-third is gobbled up by institutions as "overhead."

During your training, you are taught to write up your work in the stylized format of the scientific paper, maintaining that experiments were based on testing some question arising from existing knowledge. There is supposed to be a train of logic from formulation of the question to conclusions. This is rarely an accurate representation of reality, as I show in Chapter 5. Finally, as you become an independent researcher, you more or less recapitulate this process in writing grant applications. It is a fair question whether and how the training of scientists relates to the actual practice of science. At a minimum, it's an irony that scientists are trained to misrepresent the sequence of discovery, although applying that same process to the actual data would very likely be regarded as fraudulent.

Applying for a grant as a PI (principal investigator) when you set up your own laboratory will demand some evidence that you are capable of independent research—but you are unlikely to have established much of a track record yet. You have to attract PhD candidates and postdocs to work with you. No one is trained in how to run a laboratory, so getting off to a good start can be difficult. As the laboratory grows, more of your time is spent supervising the team, administrating within the institution, and applying for further grants, and less and less time is spent actually doing research.

Here is the irony: scientists are trained to do research, but as they become successful, further progress depends less on doing research and more on administration. Running a laboratory is somewhat like running a small business, but scientists have no business training. And writing applications for grants may take anything from 20% to 50% of their time.[3,4] If you add to this the time that researchers spend reviewing others' grant applications, the proportion of time taken up by operating the grant system becomes even more considerable. (Managing grant applications at NIH costs $110 million per year, with more than 25,000 scientists reviewing 60,000 applications and attending 2500 panel meetings.[5])

The NIH is the major source of grants in the United States. Each of its 250 Study Sections has about 24 members, all researchers in the Section's area of expertise. Grants are considered by Study Sections with appropriate expertise; of course, there is a built-in assumption here that the application will relate to a specified area of science, as opposed to a multidisciplinary approach of "big science." External reviewers can also be used to consider grant applications, but the process does not have the anonymity of peer review by journals.

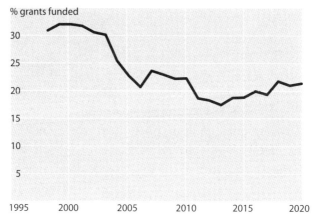

The proportion of NIH R01 grant applications that is funded decreased from 32% in 2000 to 21% in 2020.[6]

An application is scored on a scale from 1 (high) to 9 (low). Each reviewer's score is stated and the average determines the fate of the application. An application needs to get a very high score to be funded. Most grant applications are not funded, even when they are found in principle to be worth it. The consequence is that researchers have to write a lot of grant applications in order to get funded.

The grant system has its very own catch 22. One of the criteria for funding a grant is the probability of success. To demonstrate feasibility, you need to have some preliminary data to support the proposal. But you can't get the preliminary data without funding... The usual solution is to use the tail end of one grant to start work on material that will support your next grant application. In fact, it's not uncommon for a significant part of the proposed research to have been accomplished before a grant proposal is submitted, but you have to be careful not to reveal too much. Gaming the system is essential to success.

One famous example of redirection of a grant led to a Nobel Prize. Mario Capecchi at the University of Utah applied in 1980 for a grant from NIH to support three projects, two following directly from his past work, and the third for a novel project on gene targeting in mammalian cells. Reviewers praised the first two parts (the advantage of playing it safe), but did not think the new project had any chance of success. NIH funded the grant with the recommendation that Capecchi drop the third part. Instead, Capecchi put almost all his resources into the new project. It was spectacularly successful, and in 1984, the grant was enthusiastically renewed, with a recommendation starting "We are glad that you didn't follow our advice." In 2007, Capecchi shared a Nobel Prize for developing the technique to make knockout mice (in which a specific gene had been inactivated). "It was a big gamble," Capecchi recalled in 2008 about his decision to focus on gene targeting.[7]

Researchers are quick to jump into a field when a new discovery is reported. The discovery of split genes in a virus (adenovirus) in 1977, discussed in Chapter 15, turned the idea upside down that a gene must be in one piece. This sparked an immediate race to examine other viruses and to see whether this applied to genes in animal cells. By the end of 1978, at least 14 separate research groups published around 50 papers with such reports. Yet none of them could have planned to investigate this question, because the phenomenon was entirely unknown when they submitted plans for their research! Inventive uses of funding are essential.

Part of the problem for a PI in focusing on research is the bureaucracy of the university. Jose Luis Perez Velazquez tells stories of how available funds were lost because his university bureaucracy made it impossible to use them before they expired.[8] Even worse, universities may require research to be expensive (presumably because this increases the overhead that is paid to the university: on average, an institution in the United States charges a rate of 52% for its services, which means that about one-third of a grant goes to the university instead of to the laboratory[9]).

Imperial College at the University of London became infamous for threatening a professor with dismissal for failing to bring in enough money: "The metrics of a Professorial post at Imperial College include maintaining established funding in a program of research with an attributable share of research spend of £200,000 per annum."[10] Are universities now run more as businesses than as institutions engaged in the pursuit of knowledge?

When NIH grants were limited to a duration of three years, there was a vicious cycle in which halfway through the grant, you would be forced to start getting the preliminary data to write the application for the next grant. In 1986, the maximum period was increased to five years, in the hope that researchers could spend more time doing research before they wrote the next application. Even this is not a very long time: the limited duration of grants is one reason why research tends to focus on the short term rather than long term. Many historic discoveries would never have been made under such a system.

If the likelihood of discovery is proportional to researcher-hours spent at the bench, then eliminating or much reducing time spent in writing grants should lead to greater productivity in the laboratory. If researchers were to be funded on the base of past success instead of writing new proposals, any unproductive efforts that might have been caught by grant review could be compensated by the additional time spent in discovery. This is a terribly simplified argument, but it contains a nugget of truth; and certainly, wouldn't it be appropriate to use scientists for what they are trained for doing—research—rather than administration?

Some institutions do in fact function by supporting people rather than projects. HHMI (Howard Hughes Medical Institute) funds investigators on the basis of past achievements. It's impossible to establish a metric to compare the systems, because the HHMI investigators are selected precisely because of their successful track records. Moving from project-based funding to people-based funding on a wider basis might overcome many of the problems with the grant system. Of course, there remains the issue that any award still has to be evaluated, whether people-based or project-based. And it's a fair question whether a system designed for an elite can be scaled up for all science.[11]

One idea is for applications to take the form of a brief summary of projected work, accompanied by papers published from the last project. There would have to be an alternative, more traditional route, for first-time investigators to get into the system. Whether a system based on prior publications worked, or became a system for perpetuating a favored few, would depend on the rigor of assessment of published papers; so although it would undoubtedly save a great deal of time in writing proposals, it would still be subject to the familiar questions concerning peer review. And the system would have to balance the consequences of a short period of funding (encouraging short-term projects) as opposed to a long period (which would fail to stop funding of researchers whose work had run out of steam).

Attachment to the principle of peer review runs deep in the psyche of American science. When the Human Genome Project was getting under way, the Council of the American Society for Biochemistry and Molecular Biology issued a policy statement: "It is of the utmost importance that traditions of peer-reviewed research, of the sort currently funded by the NIH, not be adversely affected by efforts to map or sequence the human genome." What they had in mind was the hallowed R01 grant, officially described as, "An R01 is for mature research projects that are hypothesis-driven with strong preliminary data."

Because grant applications are a more open process compared with journal review, there have been some studies on the peer review process. They tend to show a lack of close agreement between different reviewers of the same grant.[12] It's contentious whether there is any correlation with the number of published papers or citations to them.[13]

Does peer review work better for papers (submitted after the results have been obtained) than for grants (which are prospective)? After all, there is a multiplicity of journals: negative peer reviews may prevent publication in one journal, but there is always another... But there are not so many sources for funding: negative reviews may prevent the work from being started at all. There are certainly many cases of rejections at the stage of grant application in which the scientists later found a way around and produced stunning work. Of course, these examples are outnumbered by the cases in which people were prevented from starting uninteresting work...

The low success rate of grant applications has thrown the system into disarray. "Biomedical scientists are spending far too much of their time writing and revising grant applications and far too little thinking about science and conducting experiments... . The system now favors those who can guarantee results rather than those with potentially path-breaking ideas that, by definition, cannot promise success," says one group of distinguished scientists, in an article with the strong title, "Rescuing US biomedical research from its systemic flaws."[14] They decry the "hypercompetitive culture of biomedical science."

The low funding rate discourages reviewers and applicants alike. When he took over as Director of NIH in 1993, Harold Varmus saw the problems as, "the large number of grants we're trying to review under circumstances that are made very difficult by the low success rate, the high number of resubmitted applications, the unwillingness of talented people to serve on study sections. They're making distinctions between grants that are equally excellent. It's very, very demoralizing to do that kind of reviewing."[15]

The system has reached a point of disrepute in which it has even been suggested that grants should be replaced by a lottery.[16] When the grant system was introduced, priority scores were translated into a percentile, and the cutoff for funding was at ~50%. With the cutoff now around the 80% level, the precision of priority scores is not really great enough to distinguish between percentile points. (Especially because a single adverse review is usually enough to depress the score below the cutoff.) There is more of a tendency today to look for "relevance," but basic research still has many defenders.

Would it be better to say that all proposals getting a score in, say, the top third are eligible for funding, and a lottery is used to select however many can be funded? The problem with this is that it runs counter to the value system of science, which is based on the view that data are susceptible to rational analysis. Even if it is in fact a reasonable response to the impossibility of achieving sufficient precision in assessment, it would introduce a sense of uncertainty that might discourage reviewers from participating.

The system may have outlived its usefulness, but no one has been bold enough to impose a replacement. Science is caught in a no-win situation: "the center cannot hold," as Yeats said,[17] but it's not obvious what is the way out.

Is it a sign of a wider malaise that universities often demand that faculty must bring in substantial external funds and prove their academic value by publishing in journals with high impact factors?[18] Isn't this substituting the judgment of the grant committees and the journals for their own? Wouldn't the appropriate criteria be to actually read the research articles published by the candidate to decide whether this is the type and quality of research you want to have at your institution? Ah, but that would require a lot of detailed work, and someone might have to go out on a limb...

10

FRAUD

If you're too sloppy, then you never get reproducible results, and then you never can draw any conclusions; but if you are just a little sloppy, then when you see something startling, you nail it down. So I called it the "Principle of Limited Sloppiness".[1]

Max Delbrück, 1978

Science as a culture is fundamentally chaotic, ought to be chaotic.[2]

Harold Varmus, 1993

I once gave a talk on scientific fraud at the annual meeting of a society of psychiatrists. It was an unnerving experience, because I was unsure whether the psychiatrists were really interested in my thoughts on fraud, or whether perhaps they had another agenda, such as studying the behavioral characteristics of journal editors.

It would be going far too far to say that fraud is an integral part of science, but there is certainly a case to be made that selection of data played a role in many highly significant discoveries. There is a continuing controversy as to whether and to what extent Mendel might have selected his data. The debate hinges on some highly technical arguments about how the expected ratios might have been influenced by experimental conditions, as discussed in Chapter 12.

Robert Millikan is famous for measuring the charge on an electron (he received a Nobel Prize in 1923) in what is now known as the "oil drop" experiment. Gerald Holton's reexamination of the notebooks, however, showed that the published paper mentions results from 58 oil drops out of a total of 140; the others were ignored.[3] It's now controversial whether this should be regarded as brilliant insight into which data were significant, or a case of misconduct, at least to the extent that reasons should have been given for ignoring the unused runs.[4]

Isaac Newton is regarded as one of the greatest scientists who ever lived, but his biographer Richard Westfall showed that he introduced fudge factors to get his estimate of the speed of sound and to calculate the gravitational pull of the moon.[5] A common feature in these (and other similar situations) is that there was intense competition with rivals; perhaps it was too much of a hostage to fortune to present anything but perfect data.

So here are very distinguished precedents for two discoveries—using fudge factors and selecting data—that would be regarded as questionable today. It certainly gives pause for thought that they were part of major discoveries in science. But in that era, there was the concept of a "private notebook," in which data could be recorded, not necessarily to be shown to others or used for publication. This allowed for choices that would be difficult today, when laboratory notebooks are regarded as the property of the institution and in no way private.

Ambiguity is an inevitable, if not a key, component of the way science works. Describing the work for which she obtained her Nobel Prize in 1986, Rita Levi-Montalcini said, "I began to apply what Alexander Luria, the Russian neuropsychologist, has called 'the law of disregard of negative information.' Facts that fit into a preconceived hypothesis attract attention and are singled out and remembered. Facts that are contrary to it are disregarded, treated as exceptions, and forgotten."[6] In a report on the integrity of science, the National Research Council admitted, "Selective use of research data is another area where the boundary between fabrication and creative insight may not be obvious."[7]

The nineteenth-century genius Charles Babbage, who invented the first mechanical computer—he called it the Difference Engine—wrote a book in 1830 with a distinctly modern title: *Reflections on the Decline of Science in England.*[8] Well ahead of his time, he argued that, "Scientific inquiries are more exposed than most others to the inroads of pretenders." He divided what we would now call scientific misconduct into categories that are still valid today:

- Forging—the forger records observations that he has never made.
- Trimming—clipping off little bits here and there from those observations that differ most in excess from the mean.

- Cooking—an art which [gives] ordinary observations the appearance and character of those of the highest degree of accuracy. (Under this category, he included both selective use of data and the choice of different analyses to find those that appeared most significant.)

Today we would probably call these fraud, fudging, and selection; in effect, Babbage anticipated the significant categories of misconduct. Another distinction is between fabrication (fraud) and falsification (fudging or selection). They are not a new phenomenon of the modern era, but have always been part of science.

Fraud might be considered to represent one extreme of a continuum of behavior, and, of course, different people may draw the line that constitutes fraud at different places in this continuum. The most extreme type of fraud is fabrication, simply making up data, describing experiments that were never actually performed. Perhaps because the nature of the fraudulent event is so clear, this represents most of the cases that have hit the press.

A related practice that is harder to define is to report data that were actually obtained, but that are atypical. An experiment might work once and never again, but be reported all the same. Most people would agree that it's fraud if the preponderance of data in an author's hands is different from what's reported. But most people would probably also agree that it's just sloppy science rather than fraud to fail to repeat an experiment enough times to be absolutely certain it can be reproduced.

What about reporting the same data on multiple occasions? It crosses the line to submit a paper with data that are apparently independent but that were actually reported in another paper elsewhere. In experimental science, such a practice would be difficult to conceal because original data would so evidently be the same set in each case. In clinical science, in which a follow-up study reporting a larger number of patients than a former study may be useful, duplicating data is easier to conceal.

During a period in the 1980s when fraud cases seemed to be on the rise and attracted much public attention, there was a proposal to institute some sort of audit system. To most scientists, this seemed likely to be a cure that was worse than the disease. There was a reflex that faking important findings would surely be quickly exposed anyway, and what would anyone gain by faking insignificant results? "We must recognize that 99.9999% of reports are accurate and truthful," Dan Koshland, the Editor of *Science* said in an editorial.[9]

It seems doubtful that Koshland's own research papers measured their effects to four decimal places of accuracy, but even if his assertion were true, it begs the issue of inadvertent omission of essential details. The idea of audits was to catch fraud rather than honest mistakes or incomplete reports, but given the reports of failures in attempts at reproduction, discussed in Chapter 6, I wonder if audits were such a stupid idea after all.

A minimum value for errors in the literature is given by the number of published retractions, which has increased steadily. According to reports included in the PubMed database, it was less than 0.01% until 1995, increased to 0.07% by 2006, stayed less than 0.1% until 2007, and then increased to just greater than 0.1%; it remained at this level until abruptly doubling in 2020.[10] The increase is no doubt due to greater pressure to correct the record, but, in any case, certainly significantly underestimates the number of erroneous reports.[11] At all events, the accuracy of published papers is definitely less than 99.9%!

China is the undisputed world leader in retractions. It accounts for more than 40% of the records on the Retraction Watch database, a list of retractions in scientific journals worldwide.[12] The rest of Asia accounts for well more than another 10%. The United States is at 8% and Europe is at 6%, but, of course, they account for the greatest bulk of published papers. So the *frequency* of retractions from China and Asia is very much greater than in the West (perhaps $\sim$7% for China). It's not straightforward to calculate the total frequency of retractions, but Retraction Watch has recorded more than PubMed. In the past decade, it appears to be $\sim$0.2% overall, irrespective of the field of research.[13]

Retraction is, to say the least, an imperfect mechanism. Retractions are often opaque about the reason, but fraud (including duplication as well as manipulation or fabrication of data) is the most common cause, and is the stated reason for about half of reported retractions. Of course, fraud could also be responsible for retractions that are supposedly due just to problems with data. It's followed by plagiarism[14] and errors in collecting or analyzing the data.[15] A frank statement that the data could not be reproduced accounts for only 3% of retractions.[16]

According to the data obtained by Retraction Watch, half the cases of fraud are a result of duplication of data.[17] Actually making up data is a relatively small proportion. If fraud is responsible for 50% of retractions, its frequency in papers published worldwide could be ~0.1% of publications, which seems high compared with expectations based on a traditional view of science.[18] However, given the discrepancies between different countries, the rate in the United States and Europe is very likely much less than that.[19] (Also, statistics about the overall frequency of frauds or retractions are biased by the fact that a small number of people are responsible for a disproportionately large number of retractions.[20])

A bizarre effect, which casts some doubt on the validity of judging papers by how often they are cited, is that retracted papers often continue to be cited well after the retraction, apparently because the researchers citing them do not know about the retraction. Some papers are even cited more frequently after they have been retracted than before![21] This also casts a poor light on the peer review process as well as the attention authors pay to the literature (do they actually read all the papers they cite?).

The fact is that no one really knows how accurate the scientific literature is. I don't think we could estimate it within a percentage point, let alone Koshland's four decimal places. A paper can never include all the data, and cases in which the data are not completely representative are probably quite common. Those that cross the line into wishful thinking are probably less common; those that go into fraud are probably quite rare.

It's conceivable that audits would improve the accuracy of the scientific literature, but they would have a stultifying effect on research. It remains true that significant errors will usually be corrected relatively quickly; the price for this relatively lax and informal mechanism is a good deal of waste of resources following erroneous leads.

Fraud is a solitary affair. There are no reported cases of authors collaborating in a fraud. Errors in published work, as indicated by the number of retractions, decrease as the number of authors increases. The frequency of retractions for papers with more than five authors is half the frequency for papers with up to five authors. This may be an advantage of "big science" compared with "small science." It might indicate

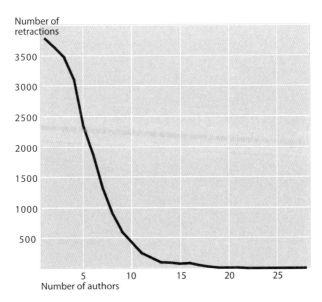

The number of retractions declines with the number of authors per paper. Fifty percent of retractions come from papers with fewer than five authors and 95% from papers with fewer than ten authors. There are zero retractions for papers with more than 28 authors. The frequency of retractions is 0.33% for papers with up to five authors, but 0.16% for papers with more than five authors.[22]

that data are scrutinized more carefully by large collaborations; more likely it indicates that the opportunity for fraud is reduced when you are part of a large team. (It could also indicate that people who are likely to commit fraud prefer to work in small groups or that frauds are more likely to be caught before publication by large groups.) I am inclined to think (admittedly without any supporting evidence) that the steep decline in retractions with increasing author numbers indicates that most retractions are in fact a result of fraud.

The presumption of innocence holds powerful sway in the world of science. Papers are reviewed on the assumption that the work is described accurately, that the authors actually did what they say they did, and that the data are representative. How else can we consider a manuscript without tearing a hole of suspicion in the fabric of science? Yet this leaves us vulnerable to any sort of deceit, not merely the extreme case of faking data, but also to any bias in selection of data or misrepresentation of unpublished data.

It's sometimes stated (without any real evidence) that fraud tends to focus on high-profile journals.[23] Even if this were true, it might be that it is simply exposed more frequently because of the high profile. I could make a guess at the frequency of fraud from my time at *Cell*, but how typical would it be of science as a whole? There is really no way to tell how frequent fraud might be in papers that are rarely cited and that don't give rise to any body of follow-up work. Most scientists would assume it is rare.

Claims surface from time to time that retractions are more common in high-impact journals. There's even been a proposal that the "retraction index" (a measure of the frequency of retraction) correlates with the impact factor of a journal.[24] This might be seen as part of a general skepticism about high-impact journals. However, I am not sure that the data set really support any firm conclusion that there's a significant correlation.[25]

Peer review is not designed to detect fraud, but one fraudulent paper submitted to *Cell* was exposed indirectly during the review process. One of the reviewers was a particularly acute researcher, who actually caused us a certain amount of trouble with his reviews. A typical review would start by acknowledging the importance of the question the paper addressed and the interest of the approach the authors had adopted. Then the reviewer would say, "I just have some technical questions," and an absolutely devastating critique would follow that entirely destroyed the results, if not the logic, of the paper. Authors, of course, would challenge the rejection by complaining that minor technical problems should not stand in the way of publishing a paper that the reviewer said was in principle of much interest.

On reviewing a paper that was the latest in the series from a well-respected laboratory, the reviewer in typical fashion remarked mildly that, although he saw no actual deficiency in the work, he was surprised to see it follow the previous paper by a margin of only two months. He calculated that the amount of work involved would have taken his laboratory at least a year. This comment prompted the senior scientist who had submitted the paper to investigate; he discovered that the experiments reported in the paper would have required the use of more reagents than the laboratory had actually purchased over the relevant period.

The paper was withdrawn. The junior scientist who had performed the experiments, still expressing his innocence, was expelled from the laboratory. On the grounds that it was not clear which of the previously published experiments were correct and which were extensions of reality, no retraction of the earlier papers was published. Some of the work, although seeming dramatic at the time, used experimental protocols with which others were successful soon after.

I have always suspected that the experiments were in fact correct, but that the junior scientist simply multiplied by some factor the number of experiments he had performed. The same paper submitted six months later might have been regarded as one of several reporting similar observations. A moral may be that it is important for forgers not to be too far ahead of their time.

This fraud, if fraud it was, fell into a pattern that was common among the (few) fraud cases I have seen close up. An ambitious junior scientist made up or extended results in the belief that he was just helping the discovery process along. He (they were all male) was too impatient to perform the experiments in full, but thought he was revealing the underlying truth. Indeed, there would no point faking something you did not believe in, because it would almost certainly soon be uncovered if it was wrong. But you would have a chance if it were in fact true…

The junior scientist involved in this affair later reappeared in the field, working in another laboratory. There was great concern among the organizers of the first meeting that he attended in his reincarnation as to whether he should be allowed to speak. A compromise was reached by giving him a brief talk very late in the evening.

A common view among the audience, however, was that each datum would have to be judged on its merits and, although needing cautious scrutiny, could be accepted as a legitimate contribution unless proven otherwise. The most surprising aspect of this attitude was that most of the people holding it agreed that the original work was suspect, but felt that the scientist should be given another chance.

Here is a dilemma. To refuse to consider any further work from this scientist is to establish a kangaroo court on the basis of the word of one senior scientist. Yet to consider his papers like any others is perhaps naive: is it appropriate to apply the usual criteria, assuming veracity of reporting, to someone thought to have faked earlier data?

Sometimes results seem too good to be true. We received a paper at *Cell* presenting results that had been reported at several meetings and had aroused suspicion. We therefore reviewed the paper rather cautiously. The only detectable inconsistency was the unusually clean nature of the data, and we suspected selective use of data rather than forgery.

We decided as a matter of policy that it is untenable to refuse publication on the grounds that other scientists had difficulty in producing results biochemically as clear as those presented in the submitted paper. Unless there's some clear statistical demonstration that the result could not be so clean, is it reasonable to penalize researchers for having unusual skill at their work? Mere suspicion, unsubstantiated by any other inconsistency or deficiency, did not seem an adequate basis for rejection. However, in due course, the work was admitted to be faked, and a retraction was duly published. If we had acted on our opinion, we would have rejected the article and averted an error, but then in other cases, we might have rejected papers that turned out to be true.

The first defense to an accusation of fraud is often that it was a unique occurrence, but the data speak quite differently. When a fraud is detected, it almost always turns out to be part of a wider pattern. It's rarely caught before any publication, and usually multiple research papers and grant applications are affected. The most likely explanation is that people who commit fraud do so habitually. The average fraudster is caught only after publishing five papers with suspect data.[26] A more cynical explanation would be that there's a certain threshold for detecting fraud, and your chances of being caught are much reduced if it really is a unique occurrence.

It is surprising how, when a fraud has been committed, those taken in by it are quick to assume that it may have been an isolated incident and that other work reported by the forger is correct (in some cases even continuing to be published). Given the susceptibility of the scientific endeavor to fraud (let alone lesser misrepresentation), the burden of proof should be on the laboratory; if any work in which a forger has had a hand cannot be judged authentic on the basis of detailed examination of the original data, it should be retracted. Is it naivete or a slowness to readjust thinking that leads collaborators to assume that they may have been taken in by an isolated incident rather than to dismiss an entire body of work?

The acid test of a paper is that if the results it presents are of any real interest, people other than the authors will attempt to build on it. They probably will not repeat the work exactly, but they will perform experiments whose assumptions are based implicitly on the previous work. If the previous work is wrong in any significant degree, some of those assumptions will be wrong, and discrepancies will emerge with the later work.

Those discrepancies may be handled in various ways. They may be dealt with by individual discussions between the researchers who were involved in both groups, and/or they may later result in the publication of a paper that disagrees with the earlier paper. This is the self-correcting mechanism of science. Sometimes it's called the self-policing mechanism of science, which is really a misnomer as it is not policing in the sense that it tries to prevent frauds or errors or even specifically to catch them. It's simply self-correcting in the sense that the intrinsic nature of the scientific process means that any significant error will result in the emergence of some discrepancy with later work.

These discrepancies with work done in different laboratories have an important place in science. Usually, they mean that experiments were not done under precisely the same conditions; when comparing the two sets of protocols, and the two sets of results, the reasons for this difference emerge. Sometimes the reasons are trivial, and sometimes they lead to the discovery of an interesting phenomenon. Science is an imperfect endeavor, and the occurrence of these discrepancies and their correction is a crucial part of the process. In some ways, this whole aspect is best encapsulated by Delbrück's principle of limited sloppiness: leave room for unexpected observations.

Sometimes a fraud is known privately well before it attracts any public attention. One factor that influences reactions to fraud is the desire not to become involved in an exposé because of the fear that some dirt may stick to the whistleblower. One researcher told me about work in a competitor's laboratory that he considered to be fraudulent, and then added that he was "too smart" to become involved by making any public comment about it. More than a year later, after several more papers had been published, doubts about the work became public from another source, and a minor scandal resulted.

In another case, a junior researcher presented some results that did not exist at a scientific meeting and admitted the situation when later confronted by the head of the laboratory; but although the situation was quite widely known on the basis of rumor, the senior researcher declined to go beyond private comment, and grant committees were faced with the problem of whether to recommend funding for the junior researcher, who established an independent laboratory. What is the appropriate response to this sort of situation?

The problem is that any systematic method of looking for fraud clashes with the value system of science. We assume that data in a manuscript provide an accurate account of what the authors did and how they did it. We view the data with skepticism, it is a vital part of the professional expertise of the scientist to look at data with skepticism. But it is the skepticism of looking for flaws in the logic, artifacts of the data, inadequacies in the controls, and alternative explanations; it would be impractical to view the paper on any basis other than to suppose that the authors have accurately described what they did.

That said, there are now some tools to reduce fraud, or at least, to increase the chances of detection. Elisabeth Bik at Stanford started a trend by comparing images in published papers to flag duplications.[27] Almost 4% of published papers contained problematic figures. This may not be representative, because 40% of the papers came from one journal (*PLoS*). However, journals are now using AI techniques to apply this sort of examination to submitted papers.[28] This approach is in its infancy, but has the potential to reduce errors in publication. Similarly, text can be checked for potential plagiarism.

A major problem in dealing with fraud, published or unpublished, was the lack of any mechanism for forcing a statement from the laboratory in which it occurred. Of three known frauds published while I was Editor of *Cell*, one was retracted in part immediately by the senior author, the second was widely known but in spite of several attempts we were unable to obtain a statement of retraction from the senior author, and the third has never been admitted, although related work was the subject of an official enquiry.

During this period, it was unclear who has jurisdiction over fraud. If we received a paper that reviewers thought might be fraudulent, we could protect ourselves by not publishing the paper, but we couldn't protect the scientific community as a whole by doing that. Viewing the paper as a privileged communication made it difficult to see how we could alert outside authorities.

The situation is easier today because since 2012 there has been an agreed code of conduct that journals should not merely reject suspect papers but should take further steps to have them investigated.[29] But it is rare to spot a fraud on the basis of inconsistencies in the data in a single paper. Unfortunately, it remains easiest to deal with fraud after it has entered the scientific literature; we can correct damage, but it's difficult to prevent.

There's an inherent paradox in this description of fraud. Science needs to be protected from fraud, because we need to be able to take a research article as a literal account of the work it describes; it would be impractical and create an atmosphere of suspicion and paranoia to do otherwise. But the self-correcting mechanism of science ensures that no substantial body of fraudulent (or even incorrect) data remains unchallenged in the literature.

Why then are we so concerned about fraud? Concern relates as much to the way we practice science as to the need to protect the community. Although we will ultimately correct any conclusions that are significantly in error, the cost of doing so can be high. In experimental research, many scientists may be set upon the wrong track, and there may be considerable delays and waste of time and resources before the situation is corrected. In clinical research, incorrect diagnoses or damaging treatments could be introduced in the period before the situation is rectified. Furthermore, the occurrence of a fraud poisons the atmosphere in which research in that area is discussed.

It's something of a myth that the self-correcting character of science leads to discovery of frauds. Analysis of known frauds suggests that the most common cause of discovery is internal, from suspicions raised in the same laboratory.[30] That's not surprising. It may be true that the frauds would subsequently have been detected by inconsistencies with work in other labs, but that of course takes longer.

The self-correcting mechanism of science is nondiscriminatory: it does not distinguish between fraud and error. All that matters is that discrepancies with subsequent work are evident. That disagreement will not necessarily lead to an overt correction or retraction of the incorrect work, but will simply render it irrelevant. Only in extreme cases does a direct attempt to replicate an earlier experiment lead to discovery of fraud.

Fraud has often been dealt with badly by the institutions in which it occurred. But it is a logical non sequitur to assume from the inability of institutions to handle sporadic cases of fraud that fraud is endemic. Perhaps it is because scientists are so convinced that fraud is so rare, that we have lacked effective mechanisms to deal with it. But any system for systematically looking for fraud is more likely to damage the scientific enterprise than to heal it.

Ability to handle allegations of fraud has no doubt been improved by the mechanisms that have been put in place since the 1980s. The problem is what to do about cases in which fraud is suspected but no one is willing to put their head above the parapet to report it.

Cases of fraud in science that became an issue in the 1980s—there were 12 prominent cases between 1974 and 1981—led to concern that science could not in fact police itself.[31] Whether or not the research record was adequately corrected, there was no doubt that institutions had failed the test of handling allegations impartially and dealing fairly with whistleblowers. This led to passage of a law including provisions that institutions should establish processes to consider allegations of fraud, and in 1989 to the establishment of the Office of Research Integrity (within the Public Health Service, but unlike its predecessor, the Office of Scientific Integrity, not part of NIH). There is now a system in place requiring all institutions accepting NIH grant money to have a defined mechanism for investigating allegations of fraud.

The ORI now publishes a record of cases in which action was taken against perpetrators of fraud. It gives the impression of a sledgehammer cracking a nut: the maximum number of cases in any year is 10, and the average is well below that.[32] Fewer than 500 cases have been handled since its inception. The surprising feature is the measured, not to say innocuous, penalties: a ban for a few years on receiving research funds.

I believe in the indivisibility of corruption: I do not believe that this is like the situation at one British university that used to have as its criterion for dismissal that junior faculty could be dismissed for immorality, assistant professors could be dismissed for persistent immorality, and full professors could be dismissed only for gross and persistent immorality. Graduating the penalty seems inadequate.[33]

Because the success of science depends so much on the way it is conducted, a fraudulent event can significantly retard progress by more than the apparent diversion to investigate the false claim. To safeguard the body of science, therefore, it's essential to hold to the principle that fraud is considered intolerable, that all allegations are rigorously investigated by outside experts, and that anyone found guilty of presenting or publishing fraudulent data is expelled from the body of science. You can't be a little bit pregnant.

I used to think that science was the last refuge of rational thought and objectivity. I am no longer so sure. Failure to handle fraud—I am tempted to say the tolerance of fraud—is understandable as a human response, but cannot be forgiven in terms of the ideals of scientific investigation. Sloppiness and chaos are one thing, but fraud is different. It places the enterprise, if not in jeopardy, at least at risk of undermining its integrity.

CHAPTER 11

POLITICS AND ETHICS

If we can make better human beings by knowing how to add genes,
why shouldn't we do it?[1]
Jim Watson, 1998

We call for a global moratorium on all clinical uses of human germline
editing—that is, changing heritable DNA (in sperm, eggs, or embryos) to
make genetically modified children.[2]
Eric Lander et al., 2019

A silomar Beach is a state park on Monterey Peninsula in California, with a conference center near the beach. There's a preserve for monarch butterflies close by. It became famous in biology as the place where scientists first considered whether there should be limits on research in molecular biology.

Physicists were brought up against the question of whether science should be unchained or whether limits should be imposed for ethical reasons by the development of the atom bomb. It's curious that although biology is more obviously, and more immediately, connected with the human condition, the question of whether ethical considerations should impose limits on the freedom to do basic research did not come up until the 1970s.

Stanford University was the epicenter for the creation and cloning of recombinant DNA and for the ensuing controversy about its safety. Several papers in 1972 and 1973 reported that by introducing cuts in two DNAs, it was possibly to cross-join the ends to create new DNA molecules (see page 40). Graduate student Janet Mertz, working with Ron Davis, used the restriction enzyme EcoRI to generate "sticky ends" (see page 41), and Mertz and Davis pointed in their paper to how this could be used to "recombine" any two DNA molecules.[3] (The term "recombination" refers back to a long history in genetics and describes the process by which genes are exchanged between the two parental copies of a chromosome when sperm or eggs are formed.)

Using a different method for generating the sticky ends, Paul Berg's laboratory inserted bacterial DNA (from *Escherichia coli*) into the monkey virus SV40.[4] The paper referred to the new molecules as "hybrid DNA." When Stan Cohen's laboratory inserted genes from a *Staphylococcus* bacterium into the workhorse bacterium *E. coli*, he also called the construct "hybrid DNA," remarking, "plasmid 'chimeras' might be more appropriate here, but we have not been able to establish a consensus among our colleagues for a definite terminology."[5] The cloning vehicle was described as a "recombinant plasmid."

The results were clearly an exciting technical development, and indeed, this was the start of DNA cloning and the biotech industry, as I show in Chapter 4. No one thought there was anything controversial about it until Janet Mertz gave a talk during a course at Cold Spring Harbor in 1971 on plans in the Berg laboratory to insert SV40 into *E. coli*, basically the reverse recombination from the earlier work.

One of the instructors in the course, Bob Pollack from Cold Spring Harbor, called Paul Berg immediately after the talk to point to the possible dangers of crossing evolutionary boundaries by putting genes from an animal virus (known to cause tumors in some animals, albeit not in humans) into a bacterium that lives in the human gut. "My first reaction was that this was absurd, I didn't see any risk to it … [but] I think I came to the conclusion that I could not say with one hundred percent assurance that this experiment would pose zero risk," Paul Berg recollects.[6] When Berg presented the results at a meeting in Sicily the following year, the reception was equally controversial, but here was focused on ethical questions about the potential for genetic engineering.[7]

As a result of these concerns, Paul Berg organized a meeting, subsequently known as Asilomar I, in 1973, in which a hundred participants (almost all from the United States) considered the dangers. They did not identify any known risks, but decided that prudence was called for.[8] After further results were presented at a Gordon Conference in the summer of 1973,[9] the participants sent a letter to the National Academy of Sciences asking it to establish a committee to consider the issues. "Several of the scientific reports at this year's Gordon Research Conference on Nucleic Acids … indicated that we presently have the technical ability to join together, covalently, DNA molecules from diverse sources," they wrote.[10] "New kinds of hybrid plasmids or viruses, with biological activity of unpredictable nature, may eventually be created."

The NAS asked Berg to convene a committee, and a group of six scientists joined him for a one-day meeting at MIT in April 1974.[11] The committee released its report three months later at the National Academy and called for a moratorium on certain experiments, in particular singling out the creation of DNAs that might confer new antibiotic resistance on bacteria, and experiments linking DNA from animal tumor viruses into a bacterial environment. They recommended an international meeting to discuss possible hazards and means of containment.[12] This was probably the first time the term "recombinant DNA" was used.

The Asilomar conference was held in February 1975 with 150 leading researchers from around the world. After a contentious three days, under pressure to show they could deal with the issues and avoid external regulation of science, the conference agreed on a report that recommended lifting the moratorium.[13] It divided experiments into categories of minimal, low, moderate, and high risk, and recommended appropriate containment facilities for each. These became enshrined in NIH regulations and effectively became mandatory for all academic research in the United States and elsewhere.

There was some collateral damage. Research in academic institutions was placed at a disadvantage compared with research in the newly emerging biotech industry, especially for universities in towns, such as Berkeley, California, or Cambridge, Massachusetts, where local regulations were passed to ban all recombinant DNA research.

Some reactions were hysterical. Mayor Vellucci of the city of Cambridge wrote to the president of the National Academy of Sciences. Referring to press reports of a "strange, orange-eyed creature" and a "hairy, nine foot creature" in the area, he asked, "I would hope …that you might check to see whether or not these 'strange creatures' (should they in fact exist) are in any way connected to recombinant DNA experiments taking place in the New England area."[14] This was worthy of the ban on the landing of flying saucers enacted in Châteauneuf-du-Pape in rural France in 1959—except that it was Cambridge, Massachusetts, home to Harvard University and MIT, in 1977.

Events in Cambridge, where the city council banned recombinant DNA research, vividly illustrated the need for science to convince the public that it can regulate itself before outsiders do so. The ban lasted for several months, effectively bringing the latest line of research to a halt at MIT and Harvard. Until then, no one had ever considered the possibility that scientific research might be subject to local regulations. Unfortunately, recent events, with the woeful response to COVID in some jurisdictions, suggest that we have not progressed very far in the past 50 years in convincing the public to respect science.

The situation in Cambridge was not helped by a dissident column in science, who believed the regulations did not take sufficient note of potential ethical problems, such as human genetic engineering, and who disrupted scientific meetings. I remember one meeting at MIT descending into chaos and having to be abandoned. There were also protests from some activists outside science who felt the issues were too important for science to be allowed to regulate itself. The majority view was expressed by Paul Berg: "I now believe that Society has more to fear from the intrusions of government in the conduct of scientific research than from recombinant DNA research itself."[15]

Science did a pretty good job of applying the same principles that govern research to a situation that was entirely novel for molecular biology: consider the data and draw logical conclusions. Any precedent in virology for needing containment facilities when working with dangerous microorganisms was not directly helpful, because there's a difference between working with a virus or bacterium that presents a known risk, and devising an experiment in which the risks have to be deduced.

The rules varied on the side of caution, but it's fair to say that the system worked. Politics fell out of the system, and, in the end, even the more extreme localities withdrew their objections to the research. Of course, it's impossible to eliminate all risk without simply halting research completely, but the exceptions have not been due to failure of the rules, but have come when laboratories have failed to follow the rules. Here, an audit system seems essential.

Twelve reporters were admitted to the Asilomar meeting, and the resulting publicity sparked the first public concern about recombinant DNA. As reporting spread beyond those who had been at the meeting, it took on a more panicky air. One of the problems in the interface between science and the public is the inadequacy (I would like to use a stronger word) of press reporting. Except for a handful of newspapers, the press has never taken science seriously.

When *Cell* published two papers in 1989 reporting the identification of the receptor for a cold virus, we thought this might be the start toward making a vaccine against the common cold.[16] (This has not turned out to be the case, of course.) We expected press reports along those lines. The work was reported as the last item on the CBS evening news—as the comic report of the evening, accompanied by an old silent movie of Charlie Chaplin with a cold. With that sort of attitude, how can you expect any rational assessment of science?

Even in the "quality" press, science can be treated with less rigor than, say, economics. In the late 1990s, when the Human Genome Project was getting under way, *The Economist* magazine reported on the project and gave an estimate for the number of human genes that was 2–3 times greater than the number expected at the time. I asked the Science Editor for the basis of this number. It turned out to be something quoted casually to him by one of the sequencers. He had not checked with anyone else or with the published literature.

When I requested that *The Economist* should publish a correction, the answer was that this was too trivial an issue to require one. If *The Economist* had made a major mistake in, say, the exchange rate of a currency, they would no doubt have regarded that as worthy of correction. (Of course, that would never have happened because it would have been regarded as serious enough to require checking.)

The COVID pandemic has undoubtedly brought science more into the public eye. The press is not solely responsible for the misinformation that has often been perpetuated. Scientists have been too willing to communicate preliminary results by press conference—but the press has been too willing to believe them. The pandemic has also revealed the extraordinary proportion of people who function on the basis of irrational belief rather than "follow the science." This should be of great concern with respect to the future of public support for science.

Recombinant DNA provoked the first of several waves of public concern about the control of science. The political situation in the United States between 1980 and 1992 essentially prevented all research on human fertilization or embryonic development. (The widely used technique of IVF was developed in Britain and no research was done on it in the United States using federal funds.) During this period, however, there was significant progress on defining embryogenesis in the mouse and on introducing genetic modifications into its germline. In 1993, Harold Varmus convened a panel at NIH to consider the implications for human embryo research. The panel recommended types of experiments that might be supported, and ran into immediate conflict with the administration. The White House issued an executive order banning NIH from supporting research involving the creation of human embryos.

The basic conflict continues to this day between those who see this as the public applying restraints to stop mad scientists from getting out of hand, against those who see it as irrational beliefs standing in the way of progress. Of course, the restrictions apply specifically to the use of U.S. government funds: they do not apply to other funding sources in the United States or to work in other countries. From time to time, American scientists have gone abroad to perform experiments they could not do in the United States. The long-term issue is that if regulations are passed that directly contradict the beliefs of the scientific community, they undercut the ability to rationally assess situations that arise subsequently. Witness the controversy as to whether NIH might have funded gain-of-function work in Wuhan.

The birth of Dolly, the first cloned sheep, in Scotland in 1997, brought the issues into sharp focus.[17] Before there could be any serious scientific consideration of the issues, an Executive Order was issued to prevent work that might lead to human cloning in the United States. Many other animals have since been cloned, and it's a common result that they are rarely completely normal; so even aside from other ethical issues, there are good scientific reasons for not attempting human cloning by this method.

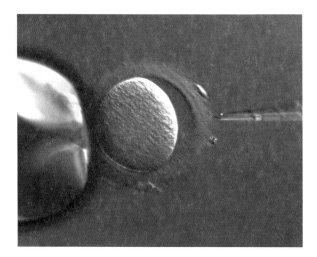

IVF (in vitro fertilization) is usually done by placing an egg in a dish with many sperm and waiting for fertilization, but a variant used in more difficult cases is to inject a sperm directly into the egg.

Dolly was made by nuclear transfer, transferring a nucleus from an adult cell into an egg, but collateral damage from the ban was to block the area of basic research from any nuclear transfer using human cells, which could have many beneficial medical effects aside from cloning, as I discuss in Chapter 20. The situation in the United States has remained difficult, with the risk of falling behind in the international arena.

More than a thousand people attended a one-day public symposium at UCLA (University of California, Los Angeles) in 1998 with the title Engineering the Human Germline. The general drift of the discussion involving some distinguished scientists was that using genetic engineering to improve the human race was more or less inevitable. This led to a flood of papers and books from participants (mostly favorable) and press reports (often critical). There was some scorn for the proposal that UNESCO should ban human genetic engineering.[18]

Concerns about genetic engineering at the time of Asilomar were overblown, and there was no plausible technique for germline engineering (using sperm or eggs) at the time of the UCLA conference, but the issues were brought back into reality with the development of CRISPR, which is so powerful a technique that it makes precise gene-editing quite easy, as discussed in Chapter 18. If a baby can be initiated by IVF, and if you have a gene editing method that can be applied to eggs or to sperm, germline editing becomes a very real prospect.

Jennifer Doudna, who shared a Nobel Prize with Emmanuelle Charpentier in 2020 for their role in CRISPR, co-opted Paul Berg and David Baltimore, who had been instrumental in Asilomar, to hold a small meeting in Napa in 2015. There was a turnaround from the enthusiasm that had been shown at UCLA. All 18 participants signed a letter to *Science* calling for a worldwide move to discourage germline editing (while noting that it was already illegal or highly regulated in many countries). They were concerned both about the ethics in principle as well as the risks of unanticipated consequences from the practice of new technology.[19] This was a fairly nuanced call for time to sort things out rather than a proposal to ban the technique.

Continuing discussion about ethics and possible routes forward were short-circuited when a Chinese researcher, He Jiankui, used CRISPR to remove the *CCR5* gene (which codes for the receptor for HIV) for the birth of twin girls.[20] The objective was to make them immune to AIDS. This clearly crossed the line into human germline editing, but to the scientists' surprise, there was no public backlash. (In China, He Jiankui was subsequently put on trial and jailed for illegal germline editing.) In the West, this time the backlash came from science.

Eric Lander, Director of the Broad Institute at MIT, where Feng Zhang was working on CRISPR and had been engaged in a patent battle with Jennifer Doudna, wrote a letter to *Nature* in 2019 calling for a moratorium, signed by 17 other scientists including Emmanuelle Charpentier and Paul Berg. Lander is a former mathematician who turned to biology, working at the Whitehead Institute at MIT before the Broad Institute was founded. He turned the Broad Institute into a major player in the Human Genome Project. He has a liking for public affairs, was Chairman of the Council of Advisors on Science and Technology in the Obama Administration, and became the Science Advisor and a member of the Cabinet in the Biden administration. (He resigned after two years following reports that he had mistreated staff.) Everyone agrees that he is very, very smart, and he is known for fighting aggressively to support his intellectual position. During the patent fight between Feng Zhang and Jennifer Doudna, he authored a review article in *Cell* that was widely criticized for seeming to credit Zhang at Doudna's expense.[21,22]

A practical argument in the call for a moratorium was that "germ-line editing is not yet safe or effective enough to justify any use in the clinic." It made the point that in almost all cases, conventional use of IVF could avoid having babies born with known damaging mutations. The letter drew a distinction between genetic correction (editing to correct a damaging mutation) and genetic enhancement (such as introducing new traits).[23]

A year later, an international commission established by the National Academy of Sciences and other august associations, issued a lengthy report arguing in effect that time was not yet right for human germ-line editing, but that the possibility should be kept open.[24] It called for the establishment of a system for oversight of HHGE (human heritable germline editing). (Lander was a member of the commission.)

The NIH was at the forefront of the early work on recombinant DNA and cloning, but then became edged out as it became impossible to work on stem cells, as described in Chapter 20. It is an enormous bureaucracy, and a prime example of the interface between science and politics.

The NIH is plural—the National Institutes of Health—and includes more than 20 separate Institutes. The Directors of NIH and the individual institutes are political appointments. Over the years, there's been an alternation of career bureaucrats and working scientists in the roles of director. There was a huge sigh of relief when Harold Varmus, a Nobel Prize winner and still a working scientist, took over the NIH in 1993 from Bernadine Healy, a career bureaucrat. Scientists had never been comfortable with the top-down planning of the Healy era, which went to confirm their belief that nonscientists (even physicians) never truly understand science. It is a mistake to think you can direct basic research along a foreseeable path.

This basic clash of misunderstanding showed itself when AIDS activists demanded that the NIH create a new institute specifically devoted to AIDS. At the time, AIDS research was mostly managed by the NIAID (National Institute of Allergy and Infectious Diseases). The problem here, from the perspective of science, is that better treatments, or a cure for AIDS, are just as likely to come from an unexpected discovery yet to be made in virology or immunology as from attempts to direct research on the basis of existing knowledge.

One difference between basic and applied research is that applied research is likely to be successful only when the basic research has been well established. In the case of AIDS, there was too much that was still not understood about the basic action of HIV.

Varmus's appointment as Director was, in fact, opposed by some activists because he had expressed skepticism about an AIDS Institute.[25] Their position could scarcely have been more shortsighted, even aside from marking an unacceptable intrusion of knee-jerk politics into science. (A compromise was reached by setting up an office to coordinate AIDS research among institutes.)

At least the AIDS activists wanted to encourage further work. Political interference is even more pernicious when it tries to stop research to pacify some segment of the population. Harold Varmus recounts in a memoir the result of a report prepared for NIH on human embryo research that laid out careful conditions under which such work would be allowed. He received a call from Leon Panetta, the White House Chief of Staff, telling him to overrule any recommendations that would allow the creation of human embryos.[26] It's bad enough that politicians should try to override the scientific process: surely it's worse when they try to make the scientists do their dirty work for them.

Belief that a free society should support the pursuit of knowledge is scarcely recent, but the case that it is a legitimate function of government, which ushered in the modern era, was made in Vannevar Bush's report in 1945 responding to President Roosevelt's call to account for the government's role in science during the Second World War.[27]

The distinction Bush drew between basic and applied research rings true today. "Basic research is performed without thought of practical ends. It results in general knowledge and an understanding of nature and its laws. This general knowledge provides the means of answering a large number of important practical problems, though it may not give a complete specific answer to any one of them. The function of applied research is to provide such complete answers. The scientist doing basic research may not be at all interested in the practical applications of his work, yet the further progress of industrial development would eventually stagnate if basic scientific research were long neglected."

It is not always easy to withstand political demands that science should be "relevant," usually meaning with regard to some specific medical situation. It's a constant battle to hold the line for supporting basic research irrespective of its apparent relevance (especially given the large amount of undistinguished research). "The public values science not for what it *is*, but for what it's *for*" is a fair summary.[28]

Ever since Vannevar Bush's clarion call—the very title of his report, *Science, the Endless Frontier*, tells you where he was coming from—to support basic research, the case for the value of freedom of enquiry has to be made anew in each generation. It's hard to put the case any better than Bush did: "As long as scientists are free to pursue the truth wherever it may lead, there will be a flow of new scientific knowledge to those who can apply it to practical problems in Government, in industry, or elsewhere."

We have sadly come a long distance away from that clear-sighted view. It may seem oxymoronic to have politics and ethics in the same chapter heading, but the effects on science are inextricably intertwined in an era in which government is the main source of funding for science. It might be unrealistic to demand that science should be unfettered in a democratic society in which the public is in effect paying for it, but it's a hard concept for biology that researchers are not free to follow the data wherever they lead. However, it is more than reasonable to demand that regulations should be based on facts and not on superstition or religion.

HISTORY

CHAPTER
12

MENDEL'S GARDEN

The unity in the developmental plan of organic life is beyond question.[1]

Gregor Mendel, 1865

I magine a conversation between Gregor Mendel and his gardener:

Gardener: "Abbot, there are more peas today."
Mendel: "Ah, what are the numbers like, are there 3 yellow for every green?"
Gardener: "There are many more yellow than green, I will count the exact numbers again…"

Science was very much an individual effort in the nineteenth century. With the development of the concept of science as a profession rather than a hobby, the PhD was introduced as the entry point into science (and other disciplines) in mid-century. Scientists worked just with students or technical assistants, and exchanges between scientists took the form of publishing research papers in what was essentially a small, closed society. Gregor Mendel's accomplishments as one of the greatest scientists of the century are all the more extraordinary for his exclusion from this system.

Working in more or less complete isolation in his monastery in Brno (now in the Czech Republic), he was so cut off that no one understood the significance of Mendel's work until it was rediscovered at the start of the twentieth century and became the foundation of modern genetics. There was simply no context with which to relate it to the nineteenth century focus on heredity and evolution. Mendel became dispirited by the lack of recognition and, after he was elected as Abbot of the monastery in 1868, gave up working in genetics. This sad history demonstrates that a discovery may have to wait until the rest of science catches up with it to be appreciated and integrated into the scientific canon. It's one of the most striking cases in scientific history of what is now known as "prematurity."[2]

There could scarcely be a greater contrast between Mendel's isolation in his monastery, and modern science, conducted by groups of researchers in full contact with the body of science. The foundations of genetics were established by one intellect working alone. A comparable achievement would be unimaginable today.

It was not even known at the time that genetic material would comprise a discrete entity. Mendel laid the basis for twentieth century genetics by showing that heredity is determined by discrete particles, which we would now call genes. This opened the way to analyze genes by mutations and to equate them with a physical property of the cell by showing they reside on chromosomes. This led to the challenge to identify the nature of the genetic material: was it protein or nucleic acid? The response to that challenge opened the modern era of biology.

Mendel started his breeding experiments with the garden pea in 1854. (The impetus was to improve the monastery's agricultural output.) As a monk, Mendel's training (for want of a better term) was religious: he had no qualification to undertake scientific investigations. But he designed a system that was decades ahead of its time.

Following seven paired features, such as yellow versus green pea color or round versus wrinkled shape, he discovered that a cross produced progeny showing only the type of one parent (yellow or round). But if the first generation progeny were crossed to make a second generation, the progeny formed a ratio of 3:1 (yellow:green or round:wrinkled).

Mendel realized this means that each plant has two copies of the hereditary factor (the gene) that determines each character. His observations would be explained if the two copies segregate independently when the next generation is formed. We don't need to worry about the exact details here; the important point is that this predicts the ratio of 3:1 in the second generation cross.

Mendel's garden in Brno. Mendel's plants were grown both outside and in a greenhouse that was later destroyed. (A new greenhouse is now being built.)

But the ratio is a *statistical average*, not an exact number. However, when Mendel published his results in 1866, they conformed *very* closely to the 3:1 ratio.[3] Suppose you toss a penny 10 times and count the number of heads and tails. On average, you expect 5 heads and 5 tails. But you would be surprised if every time you did the trial, the result gave exactly 5 heads and 5 tails. A statistical analysis of Mendel's results by the geneticist H.A. Fisher in 1936 suggested that they were simply too good, that they were something like the equivalent of getting that exact 5:5 heads:tails in many trials.

One possible explanation is that Mendel's gardener saw that the abbot was happy when the ratios were close to 3:1 and helped them along by adjusting the numbers. Another is that Mendel might have fudged the data, perhaps by simply stopping counting when the ratio was very close to 3:1, or possibly by cherry-picking the data (although this seems less likely because Mendel's record keeping was meticulous).

Probably we shall never know whether genetics is based on fudged data (although there are more recent revisionist views that in fact the reported data are not necessarily "too good").[4]

That brings us back to a critical feature in modern science. No scientist includes all the data when publishing a paper. Published work contains a selection of data that researchers believe are typical. But it is a fine line between ignoring outlier data that are atypical and selecting a set of data that conform to your expectations. Indeed, the need to demonstrate that data are truly representative is a running theme in science.

Mendel's work illustrates another attribute of the successful scientist: luck is not necessarily far apart from good judgment. All of the seven traits he studied are controlled by single genes: this is not true of all traits, and a trait controlled by multiple genes would not have shown the segregation that led to his conclusions. And each trait displayed as a discrete character: red color is dominant to white color in the garden pea—but in the snapdragon, color is quantitative not qualitative, and a cross between red and white gives pink, which would of course have been consistent with a theory that heredity depends on blending.

When the time was right, in 1900, three independent researchers rediscovered what Mendel had found 35 years earlier. The term gene was introduced soon after the rediscovery (in 1909).[5]

Mendel's work started because plant and animal breeders at the time could see that properties were inherited, but could not understand how. As genetics developed in the twentieth century after the rediscovery of Mendel's work, it seemed more often to be involved with arcane organisms (especially bacteria and fruit flies) without much immediate relevance to the human condition. By the twenty-first century, the wheel came full circle with the development of techniques to directly influence human heredity.

The concept of the gene has dominated biology ever since its rediscovery. Over the first half of the twentieth century, the focus was to identify its physical basis. The first milestone was the demonstration in 1928 that a preparation extracted from bacteria could transfer hereditary properties from one strain of bacteria to another. For the next two decades, it was thought that only protein could be sufficiently complex to carry the necessary information.

Schrödinger gave a new impetus to the debate in 1944 by publishing his book, *What Is Life?*[6] It would be hard to overestimate its influence. Schrödinger had shared the Nobel Prize in Physics in 1933 for discovering "new productive forms of atomic theory," and he analyzed genetics from the perspective of a physicist. Schrödinger started with the thought that the predicted small size of the gene (Schrödinger estimated around the order of 1000 atoms) was incompatible with its stability, because random interactions at the atomic level would render it unstable.

From this, he concluded that the gene must be an unusually large molecule (most probably a protein) forming what in the terms of physics he called an aperiodic crystal, "a regular array of repeating units in which the individual units are not all the same." He concluded that there would have to be stable bonds between its atoms. Mutations would be explained by a sudden change in state, consistent with quantum theory. Schrödinger was at pains to state that the behavior of the gene was consistent with the known laws of physics. He went on to raise the tantalizing idea that its behavior might also be determined by new, previously unknown, laws. (I recognize that I am somewhat simplifying his argument.) The challenge of explaining the gene's behavior brought many physicists into biology, creating a rush of intellectual activity that has scarcely died down yet.

The modern era started when Oswald Avery, at the Rockefeller University in New York, showed in 1944 that the material transferring hereditary properties between bacteria contained DNA. The surprise was indicated by Avery's comment, "Who could have guessed it? This type of nucleic acid has not to my knowledge been recognized in *Pneumococcus* before."[7]

Because the preparation did not consist exclusively of purified DNA—it included trace amounts of protein—it remained possible that the active principle was in fact an undetected protein. DNA was generally thought to have far too simple a structure to be the genetic material. Jim Watson recollected that when he started working on the genetics of bacteriophages (small viruses that kill bacteria) in 1947, "we were not at all sure that only the phage DNA carried genetic specificity."[8]

Skepticism about Avery's results is an interesting example of group-think based upon an incorrect assumption, an illustration that facts are not necessarily easily assimilated into the canon of science when they conflict with conventional wisdom. Ironically, one basis for the disbelief was the work of another researcher at Rockefeller University, Phoebus Levene, who developed the influential "tetranucleotide" theory.[9]

Nucleotides are the building blocks of DNA. Each nucleotide consists of a phosphate residue linked to a 5-carbon sugar, which is linked to a ringed structure called a nitrogenous base. There are only four types of nucleotide in DNA; each has a different nitrogenous base (adenine, thymine, cytosine, or guanine). The tetranucleotide theory said that DNA consists of a repeating structure that has one of each of the four nucleotides. This was so simple that it made it impossible to believe that DNA could have the variety to be the genetic material.

BACTERIOPHAGES 101

Bacteriophages (phages for short) are viruses that infect bacteria. The phage particle consists of nucleic acid wrapped in a protein coat. When the phage infects a bacterium, the particle binds to the outside of the bacterium and then injects nucleic acid into the bacterium. The nucleic acid can be DNA or RNA. A phage may multiply inside the bacterium and then kill it to release more copies of itself. An alternative for some phages is that they may insert a copy of the phage DNA into the bacterial DNA (this is called a prophage), and coexist in the bacterium instead of killing it.

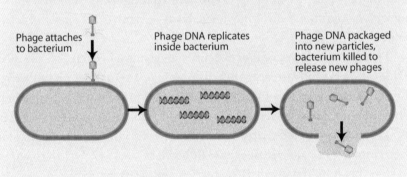

Phage attaches to bacterium

Phage DNA replicates inside bacterium

Phage DNA packaged into new particles, bacterium killed to release new phages

But in 1952, Alfred Hershey and Martha Chase at the Carnegie Institute in New York showed that the genetic material of a bacteriophage consists of DNA. It was now clear that DNA is the genetic material of microorganisms: formal proof that it is also the genetic material of higher organisms did not come until some time later.

(DNA stands for *deoxy*ribonucleic acid because the 5-carbon sugar is deoxyribose, having only an H [hydrogen] atom at a position where ribose has a hydroxyl [–OH] group. RNA is ribonucleic acid, in which the sugar is ribose. DNA is the genetic material of all cells and some viruses. RNA is the genetic material of some viruses.[10,11])

The history of genetics from Mendel to the discovery that genetic material is nucleic acid illustrates in miniature many of the forces that drive science:

* It is difficult to appreciate ideas in the absence of context. The discrete nature of the gene was difficult to understand against a background view that heredity depends on blending; DNA as the basis for heredity was difficult to understand if you believed that DNA structure was monotonous and could not provide structural information.

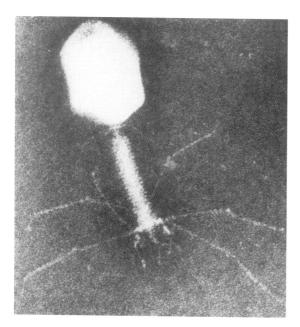

Phage T4 is an injection machine. T4 is one of the classic phages used by the "phage group" of researchers with Max Delbrück at Caltech to work out basic genetics in the 1940s and 1950s. DNA is contained in the head; the spikes on the tail attach to a bacterium, and then the tail contracts to inject the DNA into the cell.

- Luck plays a role in choosing the right features to focus on; and knowing which data to believe or disbelieve is important.

- Even if prevailing wisdom delays acceptance of new ideas, eventually the force of new data will displace incorrect theories.

At the start of the search for the genetic material, it was an open question whether it would be universal. By the last part of the twentieth century, similarities in its function across a wide range of organisms proved to be far more striking than could ever have been anticipated. Mendel's view of the "unity in the developmental plan of organic life" had proved prescient.

With DNA identified as the genetic material, the hunt was on to define its structure and to reconcile its apparent simplicity with the complexity of life.

13

THE DOUBLE HELIX

[Nucleic acids] are indispensable for life, but carry no individuality, no specificity, and ... they do not determine species specificity, nor are they carriers of the Mendelian characters.[1]

Phoebus Levene, 1916

It has not escaped our notice that the specific pairing we have postulated immediately suggests a possible copying mechanism for the genetic material.[2]

Watson and Crick, 1953

By 1951, when Jim Watson came as a postdoc to the Cavendish Laboratory in Cambridge and persuaded Francis Crick to join him in working on the structure of DNA, it was clear to them that this was the great challenge of genetics. What was the structure of DNA and how was it translated into the expression of heredity?

The *dramatis personae* were worthy of a play on Broadway: brash young interlopers, a senior scientist working on the problem but getting nowhere, an oppressed (or at least badly treated) woman, and successful senior scientists anxious to score another coup. Interactions between the participants had all the plot elements of scientific research today: ownership, priority, collaboration, and competition, except that they were missing the commercial aspects that have now become part of science.

The major group working on the problem in England was led by Maurice Wilkins at King's College, London. Their approach was to obtain crystals of DNA that could be analyzed by X-ray diffraction. Rosalind Franklin joined the group as a crystallographer, but the work was riven by continuous conflict. Wilkins thought Franklin would help him with crystallography; she thought she was leading an independent effort. They were barely speaking to one another or sharing data.

Yet the most incongruous aspect of the situation was that the structure of DNA was regarded in England as a problem reserved for the King's College group. This meant that no one else was entitled to work on it. As Jim Watson remarked in his memoir, *The Double Helix*, this sense of fair play was a uniquely British reaction.[3] "Hands-off" would not have applied anywhere else; and certainly in the world of science today, everything is regarded as fair game, to the point at which everyone may jump on a problem the moment an interesting observation is announced. This can create a dilemma as to whether to publish as fast as possible, to avoid the risk of being scooped, or whether to hold off, so as to gain some lead-time on potential competition.

The Cavendish Laboratory in Cambridge was led by Sir Lawrence Bragg, a crystallographer who had won a Nobel Prize together with his father in 1915. Within the Cavendish, in 1947, he established a small group of two scientists, Max Perutz and John Kendrew, with two assistants, to work on the crystallography of proteins.

The main competition for working out the structure of DNA came from Linus Pauling, famous for recently elucidating the α-helix in protein structure (for which he received a Nobel Prize in 1954). Bragg had been beaten to the α-helical structure of proteins by Pauling in 1951. Bragg had proposed a ribbon structure and later described his own paper as "ill-planned and abortive."[4] The history created a sense of competition with Caltech in California, where Pauling was located.

Recognized as the world's top chemist, Pauling would be formidable competition. Arguing to be allowed to work on DNA (although this was not within the Cavendish's remit), Watson said to Bragg that Linus "was far too dangerous to be allowed a second crack at DNA while people on this side of the Atlantic sat on their hands."[5] (Awareness of Pauling was sharpened by the fact that his son Peter was working at the Cavendish under John Kendrew.)

That sense of transatlantic competition was a running theme over the next two or three decades. When I was at *Nature* in the early 1970s, it was not unknown for scientists at the successor to the Cavendish Laboratory, the MRC Laboratory for Molecular Biology at Hills Road, to ask for a paper to be expedited because "the Americans are on the trail."

The MRC Unit for Research on the Molecular Structure of Biological Systems was established in cramped space in the Physics Department at the Cavendish Laboratory in the center of Cambridge. (The unit moved out of town in 1962 to become the Laboratory of Molecular Biology in Hills Road a mile to the south. The physics department moved to the new Cavendish Laboratory out to the west in 1974.)

The key development in defining the double helix as the structure of DNA was a photograph taken by Raymond Gosling, a graduate student working with Rosalind Franklin. Even though Franklin was not sharing her data, Wilkins had been given a copy of the photograph (the famous photograph "51") by Ray Gosling. It has been controversial whether he should in turn have shown the photograph to Watson. Watson and Crick had another source for the data when they saw an internal MRC report that included Franklin's unpublished work.[6] This has been a focal point ever since for issues of who owns scientific data, and what actions you are entitled to take based on unpublished information.

"The black cross of reflections which dominated the picture could arise only from a helical structure," Watson recollected in *The Double Helix*. Franklin was not immediately persuaded.[7] (One complication was that her laboratory had samples of two types of DNA: the A-form ["dry"] and the B-form ["wet"]. Franklin was mostly working on the A-form, which did not show the hallmarks of helical structure.) Subsequently, she decided that photograph 51 represented a two-helical structure. "Helices were in the air, and you would have to be either obtuse or very obstinate not to think along helical lines," Crick said in his memoir, *What Mad*

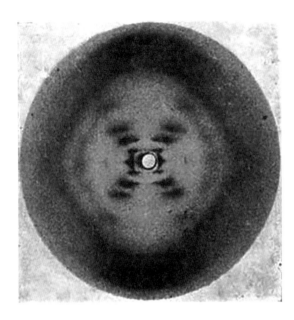

The famous photograph 51 showing the X of reflections. By Raymond Gosling, King's College London.

Pursuit, in 1988. Of course, Crick was sensitized to this issue because he was working on models for helical structures in proteins.

Perspective is everything. There are innumerable examples of failure to appreciate significant observations because of lack of context or faulty assumptions. One of England's best-known crystallographers, William Astbury, made the first efforts to model DNA. Astbury had been in the physics department at Leeds University since 1928, where he had started by working on the structure of keratins, important proteins in wool fibers (Leeds was a center of the textile industry). With his student Florence Bell, he turned to nucleic acids in 1938, and they found that DNA had a regularly repeating structure with an interval of 3.4 Å. Astbury and Bell proposed a structure for DNA like a "pile of pennies," in which nucleotides stuck out every 3.4 Å from a backbone.[8]

The data weren't good enough to proceed to any more detailed model, but Astbury's technician, Elwyn Beighton, obtained an X-ray diffraction photograph of DNA in 1951 that shows a remarkable similarity to photograph 51. Astbury does not seem to have paid this much attention: it may have seemed to him that the structure was simply too monotonous to be interesting. From his background in protein structure, he may have thought that the hereditary function of DNA would also require three-dimensional specificity.[9] It was necessary to think in terms of *information* to understand the significance. Astbury never published the photograph: if he had, Watson, Crick, and Pauling would all likely have seen it in 1952—and history might have been different. (Ironically, the first use of the term "molecular biology" is often ascribed to Astbury, in 1945.[10,11])

After early mis-steps in which Watson and Crick tried a three-helical model, and Pauling published a similar one, both with the backbone in the center, Watson and Crick realized when they saw photograph 51 and the MRC report that the backbone must be on the outside, with the two chains running in opposite directions, and the bases inside. They had recently learned of the work of Erwin Chargaff at Columbia University, published a couple of years earlier,[12] which demolished the tetranucleotide hypothesis by showing that DNA from different organisms has widely varying proportions of the four bases, but that the amount of adenine always equals thymine, and the amount of guanine always equals cytosine. Chargaff failed to progress beyond biochemical description, perhaps because he believed that "molecular biology is essentially the practice of biochemistry without a license."[13]

Watson and Crick with their model of the double helix in 1953.

Watson and Crick built a model—it stood six foot tall—from metal parts manufactured by the workshop at the laboratory. The intellectual equivalent today would be computer modeling, but it's hard not to feel that the physical model must have given a greater sense of excitement. A key feature of the model was the insight that the center of the double helix is occupied by *base pairs*, in which A (adenine) always faces T (thymine), and G (guanine) always faces C (cytosine). Base pairs are separated by 3.4 Å, and there are 10 base pairs in every turn of the helix. Watson and Crick published the model in *Nature* in 1953, together with a paper from Maurice Wilkins and two of his collaborators reporting structural analysis, and a paper from Franklin and Gosling on the crystallography.[14]

Base-pairing is one of the great concepts of modern biology, perhaps the most wide-ranging concept of all, with even more implications than Watson and Crick could have realized in 1953. It is not only crucial for replication (the duplication of DNA) and transcription (copying into RNA) but also for many other processes, most recently for the development of gene editing by CRISPR.

In a follow-up paper a few weeks later, Watson and Crick addressed some implications of the structure: "It therefore seems likely that the precise sequence of the bases is the code which carries the genetical information." Here was the concept of the genetic code, which was to drive research for the next decade.[15]

Atmosphere in the two laboratories could scarcely have been more different:

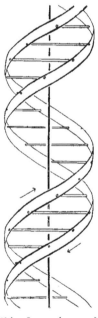

This figure is purely diagrammatic. The two ribbons symbolize the two phosphate—sugar chains, and the horizontal rods the pairs of bases holding the chains together. The vertical line marks the fibre axis

The original illustration of the structure of the double helix for the paper in Nature *in 1953 was drawn by Odile Crick.*

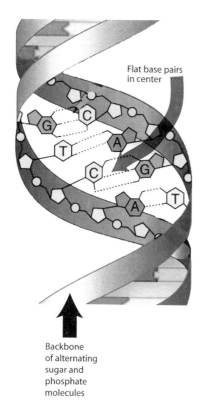

Flat base pairs
in center

Backbone
of alternating
sugar and
phosphate
molecules

*The structure of the double helix. Base pairs,
A = T and G = C, are stacked in the center,
whereas the outside is a backbone of alternating phosphate-sugar residues. Because
A can pair only with T, and G can pair only
with C, each strand predicts the sequence of
the other: they are said to be* complementary.
*Specificity depends on hydrogen bonds within
the base pairs (dotted lines). This enables the
two strands to be separated without requiring
the much larger input of energy that would be
required to break covalent bonds.*

more or less open warfare between Wilkins and Franklin at King's
College, and a rapid-fire torrent of ideas between Watson and Crick,
some right but many wrong, in Cambridge. Franklin was admittedly
prickly, but she had quite a bit to be prickly about; Wilkins was too withdrawn to resolve the situation. Watson's brashness and Crick's showy
brilliance marked the rest of their careers.

Intellectual approaches at King's and at Cambridge were at two far
extremes of the scientific method. Driven by data, Wilkins and Franklin
were both trying to "solve" the structure by accumulating enough data
from crystallography. Their approach was the same as trying to solve a
protein structure. Driven by hypotheses, Watson and Crick made inspired
guesses coming from a background of information about genetic function and biochemical properties, aided, of course, by knowledge of the
X-ray diffraction data.

Generally regarded as a brilliant insight, the structure of the double helix led to a period of outstanding discoveries in what became the new field of molecular biology. Of course, there were some holdouts: King's College Cambridge refused to give Crick a Fellowship in 1957 on the grounds that DNA research was a "flash in the pan."[16]

None of the protagonists in the race to find the structure of DNA were in any doubt about its significance. They all knew that Oswald Avery had made one of the most important discoveries of the century when he identified DNA as the "transforming principle" in 1944, but the doubts of skeptics persisted for longer than would have been reasonable if their opinions were based on the facts. Avery never received a Nobel Prize, because the Nobel committee was not convinced of its importance before he died in 1955. "It is to be regretted that he did not receive the Nobel Prize. By the time dissident voices were silenced, he had passed away," the official history records.[17] Dismissal of Avery's results went beyond any sense of "prematurity," into willful denial of facts.

Watson and Crick shared the Nobel Prize in Physiology or Medicine with Wilkins in 1962. Recent research into the archives of the Nobel Committee suggests that Wilkins was included mostly because of a recommendation from Sir Lawrence Bragg; several others were against it.[18] Rosalind Franklin was not eligible as she died in 1958, but the question has resonated ever since as to what the committee would have done had she still been alive. A long-held and widespread view is that she was never given due credit for her work. At a minimum, the failure to credit her work more fully in Watson and Crick's first paper stands in striking contrast with modern standards, which would require a full acknowledgment of any relevant unpublished work the authors knew about.

The question goes beyond whether she was discriminated against as a woman into the general issue of how to apportion credit for discoveries. What is the relative importance of obtaining the decisive data as opposed to constructing the correct theory? Indeed, the most common cause of argument about Nobel (and other) prizes is whether the senior author in charge of the group should get the credit or the person who actually did the work. The Nobel committee has usually favored the former, and has sometimes been criticized for unfairly excluding more junior researchers.

You might say that the capitalist view is to follow the money: it's the senior author who raises the funds and effectively sponsors the work. The artisanal view, perhaps, is that it's the person who actually gets their hands dirty in performing the experiment who should get the greatest share of credit. It can be difficult, and highly contentious, to disentangle intellectual from practical contributions. It is not uncommon for the wider scientific community to feel there has been an unfair omission when a Nobel Prize is awarded (a situation that is exacerbated by the limit of three laureates for a prize).

In retrospect, it seems that the structure of DNA was well and truly settled by the discovery of the double helix, but Francis Crick recollected that, "The double helical structure of DNA was thus finally confirmed only in the early 1980s. It took over twenty-five years for our model of DNA to go from being only rather plausible to being *very* plausible ... and from there to being virtually certainly correct."[19]

The model for the double helix did not emerge from testing a hypothesis as such. If there was an overriding theory, it was that the structure of DNA was a prize because—whatever it was—it would be essential to understanding heredity. I suppose you might say that a series of models (or hypotheses) were tested until one was found that stood up to the data. But in the next 25 years, there were no attempts to refute the model; rather, the efforts that turned it from plausible to virtually certain consisted of accumulating data that were consistent with the model. One of the most effective, for example, was the famous Meselson–Stahl experiment of 1958 showing that the strands of DNA separate when it replicates; this gave indirect support to the model by suggesting how the structure breaks apart.[20] The double helix is a great example of how a model becomes accepted by dint of accumulating a mass of data consistent with its assumptions. (Along the way there were proposals for some alternative structures of DNA, but these were always seen as exceptions for particular circumstances.)

What does the search for the double helix reveal about the conduct of science? Competition is an intrinsic part of science. So long as science is driven by intellectual curiosity, any interesting problem will attract multiple researchers. Any progress that makes a previously intractable problem seem more approachable is likely to bring new people into the field. The surprise in retrospect about the structure of DNA is that so few people were attracted to the problem: perhaps this was because of the long period of ambivalence as to whether DNA really was the genetic material.

By comparison with the way science is practiced today, it is surprising that Watson and Crick were able to work on the problem at all. Watson was a postdoctoral fellow and Crick had not even yet got his PhD.[21] Their accomplishment reflects the spirit of individuality in British science at the time. Today they would probably be working as part of a group under the leadership of a senior scientist, with responsibilities for specific parts of the team effort. There would be a weekly group meeting to discuss progress, and it would be unusual for anyone to go off-piste and work on an entirely different problem.

The small number of people involved is a striking difference with the practice of modern science. Jim Watson and Francis Crick formed a partnership in which each brought complementary intellectual interests. Maurice Wilkins published his report together with two workers in his laboratory. Rosalind Franklin published the X-ray diffraction photograph with student Raymond Gosling.

In that issue of *Nature*, April 25, 1953, half of the articles were from single authors, and the rest were from two or three authors, the latter being the smallest category. Look at the contrast today. In *Nature* of April 21, 2021, there were no single-author research papers; the average number of authors on a single article was 12, with the longest authorship list having 34 names.

This marks a great change in the way science is done (see p. 35). It affects both the process of creativity, as seen in the intellectual atmosphere in the laboratory, and how scientists get credit for their work. Major driving forces for the move to larger teams are the vastly greater amount of data we have today and the need for interdisciplinary research connected with increasing specialization.

The path to the double helix illustrates a feature of molecular biology that became increasingly important over future decades. Watson and Crick were able to work out the structure because they were versed in both structure and genetics. However, they nearly failed, because they started out with incorrect structures for the nitrogenous bases; a chemist at the laboratory, Jerry Donohue, put them straight about it, and then the structure fell into place. Three disciplines—genetics, chemistry, and crystallography—were needed to deduce the double helix.

Biochemists did not get there because they never saw beyond the simple monotony of only four components. Crystallographers would have got there eventually, but in the short term did not understand the implications of DNA's genetic role. A move toward a more interdisciplinary approach to questions in molecular biology has intensified steadily, as seen in collaborations between researchers with different fields of expertise, and indeed in the design of institutes, such as the Crick Institute in London, where laboratories are arranged so as to encourage interdisciplinary contacts. HHMI's (Howard Hughes Medical Institute) Janelia Campus is another example, where groups devoted to technological development are mixed with research groups on quite different topics. This is a big change from the fight for resources between different disciplines that marked the early days of molecular biology.

Molecular biology became acknowledged as a distinct discipline only slowly. Just as scientists can be slow to accept the validity of new discoveries, so they may equally question the legitimacy of new fields. Of course, disdain for other fields of science is not uncommon among researchers. Ernest Rutherford is famously quoted as saying around the start of the twentieth century that, "Physics is the only real science. The rest are just stamp collecting."[22] (That may not have been such an unfair description of biology at the time.) Ironically, Rutherford was awarded the Nobel Prize for Chemistry in 1908. He had the grace to say, "I am very startled at my metamorphosis into a chemist."[23]

The MRC Laboratory of Molecular Biology was established in Cambridge in 1962, the same year that its members (including Crick) gained Nobel Prizes in both Chemistry and Medicine or Physiology. It was (and is) one of the glories of British science, but the very definition of molecular biology was contentious well into the 1960s. A Working Group to consider the future of molecular biology in the United Kingdom as late as 1966 concluded that molecular biology did not "correspond to any real subdivision of biology."[24]

Nonetheless, the report recommended an increase in support for research in the area. A year later, the British Biochemical Society responded with its own report, in effect arguing that "biology at the molecular level" should be subsumed into biochemistry.[25] This was as much a fight for resources as an intellectual argument; the difficulty of disentangling the politics of science from the conduct of science has only increased since.

CHAPTER
14

DOGMA

My mind was, that a dogma was an idea for which there was no reasonable evidence.[1]

Francis Crick, 1975

A dogma is a provocation waiting to be overthrown. Is it an exaggeration to say that the biggest advances in science come from refuting dogmas? But that begs the question: should science have dogmas? Isn't it supposed to be based on logical deductions from facts that are subject to verification? Shouldn't science advance in steps based on existing knowledge? How can it develop a collective mindset, what in another context might be called group-think, that leads into a blind alley?

Over the decade of the 1970s, several of the major concepts resulting from the double helix model were overthrown. In almost no case did this result from a deliberate attempt to falsify a hypothesis: rather the common feature was that proceeding on the assumption the hypothesis was correct led to results inconsistent with it. Reactions varied from prolonged refusal to accept the new view to immediately embracing it.

Conventional wisdom may be tacit, but perhaps the most famous example of an overt dogma is the Central Dogma, the term that Francis Crick used in 1957 to describe the process by which DNA fulfills its hereditary functions. What he said formally was that, "once 'information' has passed into protein *it cannot get out again.*"[2] If any single phrase captures the spirit of molecular biology—indeed of the basis of life—it is the Central Dogma.

Information was defined as "the precise determination of sequence, either of bases in the nucleic acid or of amino acid residues in the protein." (The Central Dogma was accompanied by the Sequence Hypothesis,

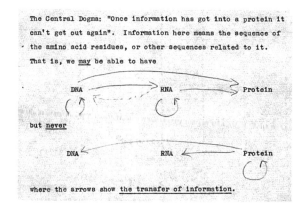

The Central Dogma: "Once information has got into a protein it can't get out again". Information here means the sequence of the amino acid residues, or other sequences related to it. That is, we may be able to have

DNA ⟶ RNA ⟶ Protein

but never

DNA ⟵ RNA ⟵ Protein

where the arrows show the transfer of information.

An unpublished note from 1956 shows Crick's thinking about the Central Dogma.

which said that the sequence of bases in DNA *codes* for the sequence of amino acids in protein.) The dogma became modified in the vernacular into the catch phrase "DNA makes RNA makes protein." This was taken to mean that DNA can be a template for directing production of RNA, and RNA can be a template for directing production of protein, but neither process is reversible.

Once DNA was defined as a double helix, the challenge that followed was to explain how a sequence made from only four types of base in DNA could specify a protein with a sequence made from 20 types of amino acids. This had two parts: what is the *genetic code* that relates the sequence of DNA to the sequence of protein; and what is the physical process by which it is executed?

The problems were solved more or less together. A series of theoretical onslaughts on the code produced no solution, but identifying the apparatus that undertook the process showed that it falls into two parts. First DNA is *transcribed* into RNA. The mechanism was foreshadowed by Watson and Crick's suggestion in their original paper that DNA could be replicated by separating the strands of the double helix so that each could act as template to assemble its complement. In the same way, one strand of DNA can serve as a template to form RNA.

The RNA that carries the genetic information is called messenger RNA (mRNA). It is *translated* into protein by an apparatus made of proteins and other classes of RNA. The genetic code runs in triplets: a group of three bases in DNA codes for a single amino acid in protein. As four

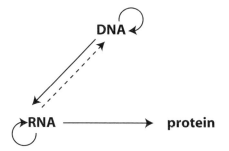

The Central Dogma today. DNA is expressed by making RNA, RNA viruses reproduce by replicating, and RNA tumor viruses make DNA by reverse transcription. RNA directs synthesis of protein but the process cannot be reversed.

types of base can give 64 triplet combinations, the code is *redundant*—that is, some amino acids are represented by multiple triplets. Some triplets code specifically for the start or end of the coding sequence.

Some of the early attempts at elucidating the genetic code showed the limits of deductive reasoning, by providing solutions that were so elegant they ought to be true—but they were wrong. (There must be a role in science for ideas that are intellectually compelling but turn out to be diametrically wrong.[3]) I am not sure to what extent it may be true in physics that elegant solutions are preferred over inelegant solutions, but biology owes more to the *ad hoc* tinkering of evolution than to elegant design.

It was easy to understand why the transfer of information to protein was unidirectional: a reverse mechanism would be a heretical contradiction of the basis of heredity. But common, one might almost say casual, acceptance of the idea that information could not pass from RNA to DNA was simply an assumption based on the way the process had been worked out. This assumption turned out to be wrong.

RNA tumor viruses convert cells into a cancerous state. Working at the University of Wisconsin, Howard Temin spent the 1960s in research suggesting that the RNA tumor viruses work by generating a DNA copy. Although the techniques were indirect and could not prove the process, part of the reason for widespread skepticism was surely based on the belief (partly from misreading of the Central Dogma) that transcription is a one-way process. Temin later described the reception of his work as, "Essentially ignored, if not derided."[4] Even only half the members of his own laboratory believed in it.[5] But in 1970, Temin was finally able to demonstrate directly that DNA can be synthesized directly from RNA.[6]

Temin had prepared the ground, but the same discovery was made more or less simultaneously by David Baltimore at MIT (Baltimore started out looking for RNA-synthesizing activity and then found DNA-synthesizing activity, perhaps demonstrating Pasteur's dictum that chance favors the prepared mind). Baltimore has never been short of confidence· that can be mistaken as an impression of arrogance, which did not serve him well some years later, when the Dingell Congressional committee investigated allegations of fraud against one of his collaborators.

In fact, Baltimore submitted his paper to *Nature* first: Temin had to scramble to get his paper in the same issue. We had some discussion at *Nature* as to whether Temin's paper should go first because of his earlier long-standing argument for the concept. In the end, we decided that the papers should be published in the order in which they were submitted, so Baltimore preceded Temin.

Together with the research articles, we published a commentary under the title "Central Dogma Reversed."[7] Perhaps this was what journalists would call "over-headlined." At any rate, it prompted a riposte from Crick, published a few weeks later, in which he pointed out that the Central Dogma had never in fact precluded transfer from RNA to DNA.[8]

In a demonstration of just how easy the experiment turned out to be, after Baltimore presented his results at the Cold Spring Harbor Symposium on Long Island in June 1970 (while the papers were under review at *Nature*), Sol Spiegelman left the meeting over the weekend, went back to his laboratory at Columbia University in Manhattan, got his students to repeat the work, and then returned to the meeting on Monday to report that he had confirmed the observations.

If I remember correctly, Spiegelman not only said that he had repeated the work, but added that it was such a pity Temin had not contacted him before he presented it at another meeting a few weeks earlier (at Houston in May), when he had met with some skepticism. (Temin was not at the Cold Spring Harbor Symposium.) "If only Howard had told me before he went (to Houston), I could have repeated the work and been there at his side to support him." But science is too competitive an endeavor for such a proposal to have been realistic. (The proposal was typical of Sol, who favored overwhelming competition with a rapid-fire publication of papers that he published in the *Proceedings of the National Academy*—as a member he could submit them for publication without peer review.)

Discussing this with David Baltimore while I was writing this book, he recollected that, "The reason that Sol could so quickly repeat the experiments is that he had all the materials at hand. So why didn't he do the experiments earlier than we did them? Because he believed in a different theory and therefore did not think about the possibility of reverse transcription."[9]

The same discovery was effectively made in two quite opposing ways: one as the result of a deliberate search for data; the other by a chance observation. It seems that it was made when its time had come.

After publication, a flood of ensuing papers led to a commentary in *Nature* a couple of months later entitled "Après Temin, le Déluge."[10] The dogma that transcription is one way had been replaced by the concept that it is accompanied by what became known as *reverse transcription*. (RNA viruses that use reverse transcription in their life cycle are now called retroviruses.)

It may seem arcane to argue about the order in which papers are presented when they are published in the same issue of a journal, but authors sometimes care about it (neither Temin nor Baltimore raised the issue). Some years later, three important papers were submitted to *Cell* reporting the same discovery. All the authors were informed that they were scheduled for the same issue.

The senior author of one of the three papers called me to explain (at some great length) that he thought his paper should go first because he had really made the first observation. I explained that our policy was that we could only go on the objective fact of the order in which the papers were submitted—otherwise, we would be drawn into making invidious judgments about priority—but if he felt the argument was compelling, he could discuss with the authors of the other papers, and I would be happy to abide by any agreement they reached.

Next day I received a call from the senior author of the paper that was scheduled to be in first place. "When X called me to say his paper should go first," he said, "I thought it really did not matter and I was going to agree, but then I realized he had got up in the middle of the night to call me [the authors were in different time zones], and so if it's that important, I demand we go first," he added with delight.

Retroviruses turned out to be important not only for demonstrating a new biological process, but practically as a cause of human disease. This required the overthrow of another dogma.

A conference on Origins of Human Cancer at Cold Spring Harbor Laboratory in 1977 reported possible involvements of DNA viruses, but generally negative results for RNA viruses. (Cold Spring Harbor Laboratory is located on the North Shore of Long Island, about 35 miles east of New York City. It is not only a research facility, but also holds a series of meetings on various topics, mostly in the summer. The Annual Symposium, devoted to a different topic each year, is one of the scientific events of the year, and the choice of subject says a good deal about what is "hot" in science at the time. Its influence is evident from the number of times it is mentioned in this book.)

The negative view of a possible role for RNA viruses in human cancer was in spite of the fact that Peyton Rous had shown in 1911 at the Rockefeller University in New York that a tumor could be transferred between chickens by a virus (now known as RSV, Rous sarcoma virus).

Cold Spring Harbor Laboratory was quite shabby in the 1970s, although today it has a snazzy campus. Participants at the Symposium gathered on the beach in the afternoons.

Previously, Vilhelm Ellermann and Oluf Bang in Denmark had shown a similar transfer of leukemia. The significance of neither work was fully appreciated: Ellermann and Bang because leukemia was not then classified as a cancer, and Rous because the transfer was thought to be an oddity of chickens. Rous did not receive his Nobel Prize until 1966.

After Ludwik Gross identified murine leukemia virus in 1951, a series of leukemia viruses was found in mice, each causing a distinctive cancer. This changed the climate of opinion and led to the search for cancer viruses in other organisms. After William Jarrett identified feline leukemia virus in 1964, subsequently showing in 1971 that it could be propagated in human cells, there was some concern as to whether it might be able to transfer to humans. Although the balance of evidence suggested such transfer was not possible, I remember that when I arrived at the National Cancer Institute in 1972, a prominent virologist who lived near me refused to have a cat in the house.

The list of species that could be infected by cancer viruses grew inexorably, with cows, sheep, and horses added before Thomas Kawakami added gibbons as the first monkeys in 1973. Bob Gallo at the National Cancer Institute found another strain of gibbon ape leukemia virus, associated with T-cell leukemia, in 1978.

In retrospect, it seems inevitable that what was true for a wide range of other animals would also be true for humans, but there were many reasons for negative opinions: a long series of reports of human cancer viruses proved to be erroneous, to the point at which they were known in the field as "human rumor viruses." This combined with the thought that they should have been easy to find if they really existed.

Opinions were so strong that in a book on cancer in 1978, John Cairns (Director of Cold Spring Harbor Laboratory before Jim Watson took over) argued that, "Certain tumor viruses ... may be offering us a way of isolating and studying the genes responsible for the cancerous state. That seems to be the real justification for putting so much effort into investigating the tumor viruses, not the vague hope that human cancer will turn out to be a virus disease."[11] Quoting this approvingly, Mike Bishop argued in a review in the *New England Journal of Medicine* in 1980 that, "few investigators would now argue that infection with a retrovirus is the sole cause of any malignant process in human beings."[12]

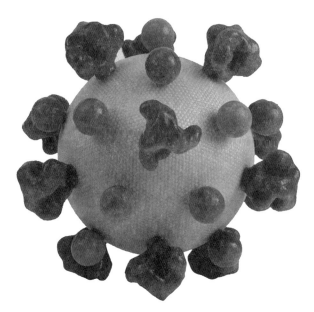

HIV is a spherical retrovirus. The surface of the sphere is a lipid membrane; the protuber-ances are the gp120 docking protein that enables the virus to enter a T cell. Inside the lipid membrane, the viral RNA and reverse transcriptase enzyme are packaged inside a shell made by the coat protein. The diameter of the virus is 100 Å.

(Ironically Mike Bishop shared the Nobel Prize with Harold Varmus in 1989 for their discovery of the cellular genes that retroviruses incorporate to become tumorigenic.)

The first human cancer virus was ultimately identified by Yorio Hinuma in 1981 in Japan, when it was called ATLV (adult T-cell leukemia virus), and subsequently by Bob Gallo in 1984, when it was called HTLV-1 (human T-cell leukemia virus).[13] The tide of opinion changed on the possible role of a virus as a result of AIDS. After AIDS was first reported in the United States in 1981, it had become epidemic by the time ATLV/HTLV-1 was isolated.

Various causes were proposed for AIDS, varying from the down-right foolish to the plausible, and resistance to the idea that a virus is the causative agent dissipated (although there was some resistance to the involvement of retroviruses, partly based on the early publication of some incorrect data). A group at the Institut Pasteur led by Luc Montagnier isolated HIV in 1983 (originally it was called LAV for lymphadenopathy-associated virus),[14] and showed that it is a lentivirus, which is a different sort of retrovirus from the leukemia viruses in the HTLV series.[15] Montagnier shared a Nobel Prize in 2008 for the discovery. One of the reasons why it's difficult to make a vaccine against retroviruses is that, because a retrovirus inserts a DNA copy of itself into the host genome, it is not adequate simply to target the virus.

The quickening pace of science is shown by the decreasing time required to overthrow dogmas. Moving from the view that hereditary material is protein to accepting the role of nucleic acids took decades. Modifying the popular view of the Central Dogma took a decade. Going from the dismissal of retroviral involvement in human disease to discovery of HIV took half a decade.

The early view that retroviruses would not be involved with human cancers was not entirely misplaced. Although the diseases with which retroviruses are associated are serious, retroviruses (fortunately) are not a common cause of human disease.

According to Thomas Kuhn in his influential book, *The Structure of Scientific Revolutions*, science advances in two ways, alternating "normal" periods (when theories are refined and advanced by accumulating further data) with "revolutionary" periods, when existing theory is refuted and replaced with a new theory. He called the latter "paradigm shifts."[16] In effect, a paradigm shift occurs as scientists struggle to fit discordant new data into existing theory and finally are forced to conclude that the theory is broken.[17]

It's implicit in Kuhn's view that major advances in science occur by finding new facts that are inconsistent with previous facts. This contrasts with the view of an earlier philosopher, Michael Polyani, who argued that, "The part played by new observations and experiment in the process of discovery is usually over-estimated. It is not so much new facts that advance science but new interpretations of known facts."[18] This is wrong, at least so far as major advances in biology are concerned, although it may explain the incremental advances between Kuhn's "revolutions."

Kuhn has sometimes been criticized for placing too much emphasis on revolution, but his view of how theories are overthrown is spot on, even if it represents an extreme case. When some known "fact" is accepted by a field, as discrepant data emerge, you see the interpretation become increasingly distorted to accommodate the "fact." Finally, it is realized that the "fact" is wrong and a new theory emerges. I say this is extreme, because sometimes, of course, the dominant "fact" is challenged immediately.

The question is, at what point do you cross the line from skepticism reflecting reasonable criticism based on known "facts," to willful refusal to accept new facts and revise conventional thinking? In retrospect, it usually seems this point comes later than it should have done. Scientists can be bound by prejudices just like anyone else.

CHAPTER

15

DOCTRINES

When you have eliminated the impossible, whatever remains, however improbable, must be the truth.[1]

Sherlock Holmes, 1890

The concept of the gene was ruled through the twentieth century by a series of changing doctrines that offer as striking an example of paradigm shifts as you could wish to find. The concept of blending was replaced by Mendelian genetics. The concept that information flows exclusively downward from DNA was at the least seriously modified by the discovery of reverse transcription. The concept that hereditary and infectious properties can be determined only by nucleic acid gave way to the recognition that proteins can be infectious. And the concept that the gene is a contiguous coding sequence of DNA was refuted by the discovery of interrupted genes.

The massive weight of evidence, accumulated in a variety of evolutionary systems, made it inappropriate to call it a dogma that only nucleic acid can be genetic material: it was more like a law of Nature, comparable to the effects of gravity. The twentieth century's most influential philosopher of science, Sir Karl Popper, had long argued that a hypothesis can never be proven, only disproven, but admitted in later works that he had underestimated the significance of accumulating evidence consistent with a hypothesis.[2] At all events, the hereditary role of nucleic acid became the unquestioned and unquestionable paradigm of genetics.

So when Stanley Prusiner of UCSF (University of California San Francisco) started publishing a series of papers attributing infectious properties to protein, it was nothing less than heresy. It was doctrine that all infectious agents had genomes of nucleic acid. At first, the papers did not so much attract hostility as simply indifference, because the conclusion was impossible to assimilate. The most significant papers were submitted to *Cell*, and although I could not really say that we had to overrule the reviewers in order to publish them, it would be fair to say that the reviewers' assent to publication was at best reluctant.

Our guiding rule at *Cell* was that, if a paper was interesting in principle, we would publish it unless we could find a flaw in its logic or data. Any other approach falls prey to prejudice or value judgment, and to my mind is antithetical to the basic rule of science that research is judged solely on the data.

Prusiner spent a long time excluding the possibility that the infectious agent was a nucleic acid. When he decided it must be protein, in 1983, he called it a *prion*, for proteinaceous infectious particle.[4] Claims that protein was the infectious agent met criticism that the purification procedure had not excluded the possibility that the real agent was an undetected nucleic acid.

This was an ironic reversal of the arguments when DNA was identified as the genetic material: that the real agent might have been an undetected protein in the preparation! There was a certain similarity in the critics' reactions in both cases, in believing that the lack of an obvious mechanism to explain the results meant they must be wrong.

The protein on which Prusiner worked, called PrP, turned out to be responsible for a series of diseases including scrapie in sheep, CWD (chronic wasting disease) in deer, BSE (bovine spongiform encephalopathy) in cattle (known more casually as mad cow disease), and CJD (Creutzfeldt–Jakob disease) in humans. Prions are transmissible and can sometimes cross species barriers, which is why eating infected beef could cause the so-called mad cow disease.

The key idea is that the normal form of the protein, PrP^C, can change into an abnormal form, PrP^{Sc}. The two forms have different structures. At the same time, a related phenomenon was found in yeast, extending the principle that protein structure can be independently inherited.[5] Prusiner won the Nobel Prize in Physiology or Medicine in 1997 (and somewhat unusually for science these days, was the sole recipient).

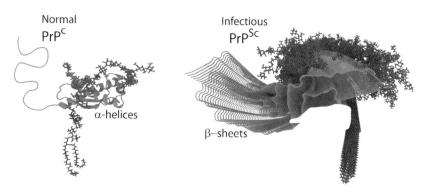

Normal
PrP^C

Infectious
PrP^{Sc}

α-helices

β-sheets

A proposal for prion structure shows how the normal protein changes to the infectious form by seeding the conversion of α-helices into β-amyloid sheets.[3]

PrP^C "normal" protein exists as a single unit, with a central core consisting of two α-helices (the α-helix is a common feature of protein structure). The infectious PrP^{Sc} form consists of aggregates in which the α-helices have been converted to flat sheets (technically known as β-amyloid fibers). The sheets are sticky, and cause new PrP proteins to convert their α-helices into β sheets that extend the structure. The exact structure of the sheets varies with the strain of prion, and is effectively a heritable property.[6]

The behavior of prions actually demolishes not one but two dogmas: first, that only nucleic acid can be infectious; second, that protein structure follows ineluctably from sequence. The idea that PrP could exist in an alternative structure that it was able to enforce on other copies was a break with the principle that the structure of a protein is determined solely by its sequence (which in turn is determined by the sequence of the gene that codes it). In effect, PrP can have alternative structures with distinct infectious properties, even though the different protein forms have the same sequence.

When a new molecule of PrP is synthesized, it takes up its default structure of Prp^C if there is no PrP^{Sc} around, but it will take up a different structure in the presence of a PrP^{Sc} template. Structure is determined not just by sequence, but also by the environment within the cell.

Christian Anfinsen had shown in a classic study in the 1950s that after the protein ribonuclease is denatured (treated to cause it to lose its structure), it can refold into its original structure. The conclusion that the sequence of a protein carries the information necessary for it to fold into the correct structure won Anfinsen the Nobel Prize in 1972.

We now know that the system is not quite so simple, and that accessory proteins (called chaperones) assist and influence folding. Even so, there was no precedent for the ability of a particular conformation of a protein to impose its own structure on newly synthesized molecules instead of the one they would otherwise have folded into.

When I recently asked Stan Prusiner when was the tipping point from disbelief in a proteinaceous infective agent to accepting the possibility, I expected him to indicate some time between the first paper in 1983 and his Nobel Prize in 1997. (I remember the tide turning some time in the early 1990s; after Prusiner gave a talk at a conference, one scientist, who had reviewed some of his submissions to *Cell* with

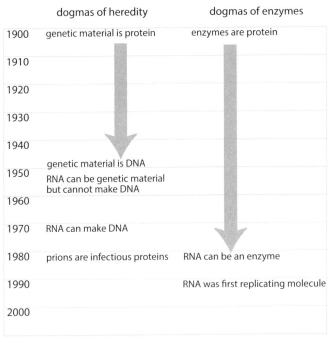

Timeline for dogmas of the twentieth century. A dogma that only proteins had enough complexity to account for both heredity and enzyme activities filled the first half of the century. The transition to the dogma that DNA is the genetic material took a decade from 1944. The time shortened for subsequent paradigm shifts recognizing new roles for RNA and proteins.

skepticism, turned to me and said quietly, "that man is going on a trip to Stockholm"—he meant that Prusiner would get a Nobel Prize).

Stan laughs in response to my question and says, "Well it hasn't happened yet," going on to explain that he is now working on finding drugs to treat Alzheimer's, basing his approach on similarities between the amyloid fibers found in Alzheimer's and the structure of prions. "Anyone who mentions 'prion' in a grant application for Alzheimer's research never gets funded," he says. He thinks this is partly personal. "People want to say they discovered it, not that I did," he says.[7]

Indeed, you certainly need a certain perseverance to spend years in the wilderness advocating a theory that is widely rejected, perhaps a larger than life quality that can be perceived as brashness. Prusiner certainly had to exhibit some forcefulness to overcome years of skeptical reception, which sometimes broke out into outright hostility. When persistence irritates others, a theory may even be rejected as a reaction to the personality of the author.

Even the Nobel Prize did not end the criticism that the *real* agent of scrapie might be a small virus that had been undetected. In a demonstration of the extreme hostility to the idea that scrapie is caused by protein, Prusiner describes in a memoir a series of press reports (aided by commentary from scientists) saying how embarrassed the Nobel committee would be when the virus was discovered.[8] Some doctrines die really hard.

The work on prions established a new principle, effectively that a protein structure can be infectious. This was certainly not conceived as a possibility in the Central Dogma, although it is not at odds with it. Publication of the series of papers on prions in *Cell* and elsewhere, in spite of the fact that there was no way to understand the results in terms of conventional wisdom, is an example of the way science should work: follow the data.

Looking back when there's a paradigm shift, you can often see how everyone was blinded by the same assumptions that permeated the field. The roles of nucleic acids and proteins were stereotyped for much of the twentieth century. For the first half of the century (and beyond!) no one thought nucleic acids could be the genetic material. And no one ever questioned the thought that only proteins could be enzymes.

Enzymes are responsible for all the metabolic (and other) reactions of living organisms; they work by catalyzing reactions, which means they increase the rate without themselves being changed. An enzyme creates a catalytic center within its structure that brings the two reacting species together. It seemed obvious this requires the complex three-dimensional structure that only proteins can provide.

Trying to purify the enzymes that modified the structures of certain RNA molecules, Sidney Altman at Yale and Tom Cech at the University of Colorado found that they could not separate the activity from the RNAs.[9,10] With apologies to Sherlock Holmes, it follows that if there is no extrinsic factor, the property must be intrinsic. Altman and Cech shared the Nobel Prize for Chemistry in 1989.

Cech was working on the processing of an RNA in *Tetrahymena*, a model unicellular organism, only two years into his first faculty position, when he found that the RNA undertook its own reaction. He described the problem in the original paper. "We hoped to use the unspliced

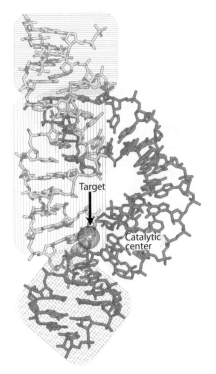

Named for its resemblance to a hammerhead shark, the hammerhead ribozyme is one of the best-characterized ribozyme structures. Just like an enzyme, it has a distinct structure that creates a catalytic site. Its four domains are indicated by different cross-hatching. The catalytic site is adjacent to the circle that marks the target for cleavage.[11]

pre-rRNA ... as a substrate to assay for splicing activity in nuclear extracts. This approach proved to be impossible, however, because [the activity] occurred when the isolated RNA was incubated." (The splicing activity is the ability to break the RNA and join the new ends together.) This was an unusually clear account of how the observation was really made.[12]

Tom recollects how the transition in his thinking occurred. "It took a year to a year and a half to discover there was no protein. I like everyone else was in the dogma that all reactions were catalyzed by enzymes and that we must somehow be missing a protein, for example, one that was bound to the RNA. We kept sticking to that although the data were to the contrary. We could boil the RNA in SDS [a procedure that inactivates all proteins] and it still worked. So we reversed the hypothesis and asked how we could prove there was no protein contaminant."[13]

Cech later coined the term ribozyme to describe an RNA with catalytic activity.[14] "We had a lab contest to find a name for the activity. Many of the names were quite specific to *Tetrahymena*. One fellow came up with ribozyme—it was pretty gutsy to name a whole class when there was only one example. But if it turned out to be unique, no one would care and it wouldn't matter. There are now more than 5,000 systems that depend on RNA catalysis."

Catalytic activity does require a specific three-dimensional structure, but this can form within the RNA. Many examples of ribozymes that can break and make bonds in their own structure or in other, target RNAs, are now seen as an integral part of the apparatus that executes the genetic code to make proteins. The range of catalytic activities associated with RNA supports the concept of the RNA World Hypothesis, the idea that life may have originated with RNA rather than DNA.[15]

During the decade of the eighties, prevailing dogmas for both nucleic acids and proteins were overthrown. Prions are an example of proteins doing something only nucleic acids were thought able to do. Ribozymes are an example of RNA doing something only proteins were thought able to do. In both cases, research efforts were directed to finding the infectious agent that was predicted by existing dogma, in effect to confirming the current hypothesis; the model broke when this proved impossible. Failure to find an external factor that undertook the process meant that it must be intrinsic.

Why was there great resistance to the idea that prions could be infectious, but little resistance to the idea that ribozymes could be catalytic? They were equally heretical, and in a sense, depend on the same sort of insight: that these unexpected properties are associated with acquiring a specific three-dimensional conformation. Part of the answer is that the mechanism of RNA catalysis was explained at the outset, whereas the mechanism for the prion's change in conformation is still not really well understood.

When I asked Tom why his results were so readily accepted, he said that, "I'd like to think it was because we were so rigorous in our work, but a lot was because splicing is a chemical reaction and saw bonds being made. When you see something like that, it gives a handle on the system that is undeniable—bonds don't just form like that—whereas the prion was an aggregate in a dirty system, for example from a brain, and there was no real chemistry."

Tom's chemical approach and intense focus on the question at hand has continued to this day. I happened to speak with him the day his Nobel Prize was announced, about a paper he had submitted to *Cell*, and I remember that he brushed aside my congratulations to get to the matter at hand. "My job today is to get this paper published in *Cell*," he said.

The question of whether and how new data are assimilated into the canon of science is sharper in physics, in which there is a clearer relationship between experiment and theory. Sir Arthur Eddington, who obtained key data to support Einstein's theory of relativity, expressed a much-quoted view in 1934. "It is a good rule not to put overmuch confidence in a theory until it has been confirmed by observation... . It is also a good rule not to put overmuch confidence in the observational results that are put forward until they have been confirmed by theory."[16] The second part is not always taken seriously, but although in formal terms it seems problematic, it captures an essential truth about the scientific attitude: observations may not be accepted until there is some theoretical framework that makes sense of them.

Before the genetic code was broken, the principle was established that a gene is *colinear* with its protein product. Basically, this means that the continuous sequence of the DNA of the gene codes for the sequence of the protein. For the next 20 years, with research focused on microorganisms, this appeared to be a universal truth. But when research turned to higher organisms, it turned out to be the exception rather than the rule.

Using independent approaches, Rich Roberts and his collaborators at Cold Spring Harbor, and Phil Sharp and Susan Berget at MIT, showed in 1977 that a gene is not necessarily continuous.[17,18] An interrupted gene consists of a series of *exons* that have sequences coding for protein, alternating with *introns* that have no coding function. Colinearity between the nucleic acid and protein sequences is created by *splicing* the sequences of the exons together when messenger RNA is generated. (The introns are thrown away.) For the most part, exons are joined in the order in which they occur in the gene, but there are exceptions when different sequences can be created by alternative splicing events. Rich Roberts and Phil Sharp shared a Nobel Prize in 1993.

Rich Roberts was one of the British contingent working at Cold Spring Harbor—Jim Watson always had a liking for hiring Brits. What everyone remarks about him is his laser-like focus on the problem at hand. He was trying to determine the sequence of adenovirus, using an approach that started with breaking the virus into defined fragments. Collaborating with electron microscopists Louise Chow and Tom Broker, he found a discrepancy between the structures of the viral fragments and their RNA products.

Phil Sharp comes from a rural background in Kentucky—you can still just hear it in his accent. He and Susan Berget were mapping adenovirus and trying to work out why its RNA products were much shorter than expected when they found the discrepancies leading to the discovery of split genes. Phil submitted a paper immediately to the *Proceedings of the National Academy*, because he did not want to risk being held up by the review process at a more conventional journal. He's a strong believer in the view that once data are out there, they are fair game for everyone. Phil is wickedly smart, with an exceptionally keen eye for discrepancies. He was later one of the founders of the biotech company, Biogen.

The discovery of splicing was as great a break with conventional wisdom as the discovery of reverse transcription or prions, but it was accepted immediately. The level of surprise was indicated by the title of Roberts' paper, which started "An amazing sequence arrangement…"

(I usually discouraged authors from using terms such as "surprising" or "astonishing" in research papers on the grounds that such descriptions may tell us more about the perspicacity, or lack thereof, of the authors, than about the work, which should speak for itself, but Rich was adamant, so this was one of the few papers that made it into *Cell* with such a title. "Look, this really *is* amazing," Rich said, or something like that. My other bugaboo was authors stating "the experiments were performed carefully." What would be the alternative?).

Both Roberts and Sharp were working on adenovirus, but many laboratories rapidly jumped into a race to test genes in higher organisms. This says a lot about how science really works. Most laboratories function on grants for working on specific projects. No one could have a grant to work on interrupted genes before they were discovered, but immediately postdocs were pulled off their existing projects in laboratories all over the world and transferred to work on the new discovery. Science can be pretty versatile when the significance of a paradigm shift is appreciated.

Why was there such a difference between the reluctance to accept reverse transcription and the rush to jump on the bandwagon of interrupted genes? Phil Sharp says that, "The early experiments on which [reverse transcription was] based were not definitive. In fact definitive experiments did not appear for decades.... In the case of the discovery of split genes/RNA splicing, the early science hinting at something quite different between the pathway of synthesis of mRNA in bacteria and that in mammalian cells, was the hnRNA findings. [mRNA is messenger RNA, the RNA copy of the coding sequence of a gene. hnRNA stands for heterogeneous RNA, a set of large RNAs found in the nucleus.] The definitive experiment by electron microscopy revealed the process of RNA splicing and established the relationship between hnRNA and cytoplasmic mRNA.... Within 6 months, everyone interested in the topic was convinced of its importance."[19]

A significant point here is that if you're going to propose a heretical conclusion, it's best to have strong evidence right at the start. One of the reasons Howard Temin struggled in the wilderness for so long was that his first proposal for reverse transcription, as discussed in the previous chapter, was based on rather weak, indirect evidence, little more than an inference. This led to the idea being dismissed, and the threshold for providing convincing data was increased by the need to overcome past skepticism.

In one sense, the discovery of interrupted genes explained a long-standing puzzle. Messenger RNA carrying the sequence from DNA to protein in bacteria was known to be short and unstable. Its counterpart in mammalian cells was the much longer hnRNA, for which no function was known at the time. The need to "splice" the product of transcription in order to generate messenger RNA suggested that hnRNA might be the immediate RNA product made from the gene and the intermediate that was processed to give the mRNA. So even though the discovery was entirely unexpected, it was possible to place it in the context of explaining puzzling past observations rather than to see it as a refutation of existing ideas. The electron microscopy from the Cold Spring Harbor group was so striking and immediately convincing that we put it on the cover of *Cell*.

Indeed, Rich Roberts says that, "the day we obtained the first EM image, all the previous doubters were convinced," and adds that "[many people] had data that they had not been able to understand before. This also explains why so many people were rushing to get a 1977 publication date and claim they had really discovered it!"

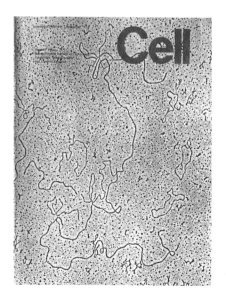

The cover of Cell, *from September 1977 with the electron micrograph showing splicing of adenovirus. Loops extending from the main axis are the introns that are excluded from the messenger RNA.*

Francis Crick summarized the effects of the discovery two years later. "There can be no denying that the discovery of splicing has given our ideas a good shake... . What is remarkable is that the possibility of splicing had not been seriously considered before it was forced upon us by the experimental facts... . Lacking evidence, we had become overconfident in the generality of some of our basic ideas."[20]

The moral of the shifts in perspective following the discovery of the double helix is that devotion to what seem to be "facts" can actually impede science. My own view is that, using the term "dogma" in its proper meaning, it is when theory becomes dogma that the progress of science is threatened. Biology is a difficult science in which to hold absolute beliefs—only nucleic acids can be infectious, only proteins can be enzymes—because it is intrinsically messy. Skepticism is a protection against falling into dogma, but misapplied can become a danger.

A striking feature of the great discoveries of the decade is the relatively small number of people directly involved: most of the research papers had three or fewer authors; only a handful had as many as six. This was the apogee of question-driven science.

The two decades following the discovery of the double helix were a period of unparalleled ferment and revelation, although in 1968, Gunther Stent, a former member of the phage group under Max Delbrück, published an influential review, "That Was the Molecular Biology That Was,"[21] which was widely taken as a lament for the decline, if not the fall, of molecular biology.[22] Stent had been right there at the beginning when Delbrück, a physicist turned biologist, turned to bacteriophages [bacterial viruses] as a model system, and the phage group grew up around him in the 1940s and 1950s.

Stent pointed to some confusion in definitions of what constitutes molecular biology, with a conflict between the structural school (believing that biology can be understood only in terms of the three-dimensional structure of its elements) and, on the other side, the informational school (believing that the key to the subject was the storage and expression of information).

Perhaps not surprisingly, given his background, he concluded that the influence of the structural school had been "nonrevolutionary," because they were following the general precept that physics could contribute to biology, whereas "some of the early informational molecular biologists were motivated by the fantastic and wholly unconventional notion that biology might make significant contributions to physics."

So he divided the history of molecular biology (meaning the informational school) into three phases. The "romantic phase" was driven by the quest for the physical basis of the gene. His view of this phase was that the conclusions drawn from the experiments were right, but the speculations built upon them were wrong, concluding in the idea that protein is the genetic material. The "dogmatic phase" started with the discovery of the double helix and lasted for a decade, until it became clear that the central dogma in effect refuted any notion that new laws of physics might be needed to explain heredity. For the final, "academic phase," "what remained now was the need to iron out the details."

There was one exception: "There now seems to remain only one major frontier of biological inquiry for which reasonable molecular mechanisms still cannot be even imagined: the higher nervous system." But he was depressed by the idea that searching for a molecular basis for consciousness might be a waste of time, as it would end up simply characterizing the same sorts of reactions that were already known.

The period since then has been marked by two conflicting trends. The refusal of romanticism to die is shown by epigenetics, a field in which hereditary characteristics are ascribed to factors other than the sequence of DNA. As a rational field of study, it follows reductionist principles in defining the molecules that are responsible, but sometimes it descends into mysticism (especially in considering effects of the environment on heredity, as I show in Chapter 19). And by way of complete contrast, the project to sequence the entire human genome follows a reductionist belief that everything can be explained by the sequence of our DNA.

Scientists are perpetually terrified of finding they have discovered everything and there is nothing left to do. At the very start of the twentieth century, in 1900, Lord Kelvin, who discovered the electron, is supposed to have said, "There is nothing new to be discovered in physics now. All that remains is more and more precise measurement." Albert Michelson, of the Michelson–Morley experiment that described light as a waveform, is supposed to have said in 1894, "Most of the grand unifying principles have been firmly established ... the future truths of physical science are to be looked for in the sixth place of decimals."

Although neither of them actually said something so pessimistic,[23] the point is that the (mis)quotations were widely circulated as examples of current thought. The basic anxiety of scientists, that things will not turn out to be interesting or profound, was summarized by physicist Niels Bohr, who used to say that, "it was the task of science to reduce deep truths to trivialities."[24]

So where does this leave molecular biology? Is Stent's elegiac lament, that we have come to the end of molecular biology, coming true forty years later? Or are there still frontiers to vanquish that will yield new insights into the human condition?

LANDMARKS

CHAPTER

16

MAPPING

The story that Francis came into the pub and said, "I have found the secret of life," I made that up of course.[1]

Jim Watson, 2017

Nothing could better epitomize the change in science between the middle and end of the twentieth century than the research projects Jim Watson was involved in at the start and then toward the end of his career. In 1953, he discovered the structure of DNA with Francis Crick in a classic collaboration between two individuals. By 1990, he was helping to establish the Human Genome Project, a massive collaborative endeavor, in which authorship on the paper published in 2001 was attributed to the "International Human Genome Sequencing Consortium" and a (partial!) list of authors was confined to a footnote that filled a whole page.

The transition from individual efforts to large groups was nothing if not controversial, but did not happen instantly. As molecular biology grew from a field in which papers could be written based on extrapolation, if not speculation, from a small number of facts, to a field driven by techniques, mastery of technique became more specialized, and groups were needed to bring together researchers with expertise in different areas. This took place more or less in the 1980s.

The advance of science is always limited by techniques, but the development of sequencing is one of the most striking examples of technique driving discovery. Sequencing started painfully slowly, not much more than an amino acid at a time for proteins in the 1940s, and then a few bases at a time for nucleic acids in the 1960s. The move to large-scale sequencing of DNA in the 1990s was a revolution, not merely for the exponential increase in knowledge, but for a change in how research is done in biology. Of course, making this work also required advances in ancillary techniques for manipulating DNA. There is no better demonstration that the development of methods can be as powerful a driving force in science as new ideas.

Just as the scale of sequencing has changed enormously, so research has moved from comparing individual experiments with their controls to scrutinizing vast collections of data for correlations, as discussed in Chapter 2. The two approaches exist in parallel, but this is a sea change in biology.

One of the most important consequences of the discovery that DNA is the genetic material, and that it has the form of a double helix, was a transition from supposing that there would be some sort of *structural* relationship between the genetic material and the components of the cell. Instead we think in terms of *information*: the sequence of DNA codes for sequences of amino acids, which in turn determine the structures and properties of proteins. So the sequence of DNA is the ultimate definition of a species.

The state of knowledge about the human genome when the double helix of DNA was discovered was so primitive that the number of chromosomes was not even known correctly. Theophilus Painter had identified the number as 48 in 1923, consisting of two sets of 23 autosomes plus the pair of sex chromosomes, XX in female and XY in male.[2] (Until this time, there was controversy as to whether males possess a Y chromosome or simply have only one X chromosome. The autosomes are all the other, nonsex, chromosomes.)

The number 48 became set in stone, although it was in fact at the higher end of a range of results from 45 to 48. If a lower number was found in later studies, the general assumption was that some of the chromosomes must have been missed. It was only in 1956 that improved techniques led Albert Levan, working in Lund (Sweden) to correct the number to 46 (22 autosomes plus the sex pair).[3] The conclusion of his paper referred to colleagues who had also recently found 46 chromosomes, but abandoned the study "because they were unable to find all the 48 chromosomes in their material." The paper also pointed out that many of the published data actually conformed better to 46 than 48. Researchers had been blindsided by the known "fact" of 48 chromosomes.

Distance in cM Locations of genes

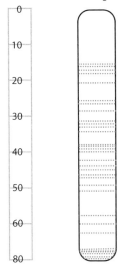

The linkage map of mouse chromosome 10 in 1990 contained about 30 genes (dotted lines). Distance between them was measured in centi-Morgans (percent recombination). The total linkage length of the chromosome is 77.9 cM; the physical length is 131 Mb. Gaps between genes varied between 0.9 and 15 cM, so the closest mapped genes average ~1.5 Mb apart, and the longest area without any mapped genes is ~10–15 Mb.[4] The map contains only ~3% of the approximately 900 genes on the chromosome.[5]

By the start of the 1960s, it was widely accepted that DNA was the genetic material, that in humans it was contained in the set of 22 autosomes plus the sex chromosomes, and that a mutation in a gene resulted in a change in the sequence of the protein it specified. Vernon Ingram had shown in 1957 that the mutation creating sickle cell anemia results in a change in the sequence of the blood protein, hemoglobin.[6] But with the sole exception of genes on the X chromosome, it was impossible to identify which chromosome might carry any specific gene. (The X chromosome is different because sons obtain an X chromosome only from their mothers, which gives a distinctive pattern of inheritance. The gene for color blindness was the first example.)

The first genetic maps were based on *linkage* between mutants. When two genes are on the same chromosome, instead of obeying Mendel's law of segregation, they tend to stay together. It's possible to use this to make a *linkage map*, giving the relative distances between genes.

A linkage map of the mouse in the early 1990s had about 700 mutant loci,[8] just over 3% of the number of genes we now know the mouse has. But even this level is not practical in humans, because there are neither enough mutants nor situations in which inheritance can be followed in families.

LINKAGE 101

———————————————◼———————————————

A linkage map is a matter of statistics. If two genes separate only 10% of the time, they are closer together than if they separate 20% of the time. The tendency to separate is called the *recombination frequency*, measured as a percentage." Once you have the recombination frequencies for a set of genes, you can place them in an order. Only genes on the same chromosome show linkage. A greater recombination frequency means genes are farther apart. The limit for recombination frequency is 50%; when two genes separate 50% of the time, they are effectively obeying Mendel's law and may be on different chromosomes (or far apart on the same chromosome). When two genes have 0% linkage, they are essentially so close together that they cannot be separated. In organisms such as the fruit fly, in which many mutants can be found or induced, a linkage map can be extended all along a chromosome. A linkage map has the limitation that it is based only on those mutants having a visible or measurable effect: so the approach is more useful for microorganisms or lower eukaryotes such as fungi than for higher eukaryotes, for which it's difficult to get enough mutants.

A breakthrough to the modern era came in 1980 when David Botstein at MIT proposed that differences in the sequence of DNA between individuals revealed by a certain class of enzymes could be used instead of mutants to make a linkage map.[9] Here is another example of a discovery in one field, one which would never have been funded had the criterion been relevance to human welfare, jumping across to create new insights in a completely distant area.

In the 1960s, Werner Arber at the Biozentrum in Basel showed that bacteria have systems to protect themselves against invasion by bacteriophages (viruses). The bacteria produce a *restriction enzyme*, which recognizes certain sequences in the viral DNA as targets for attack; the virus is inactivated when its DNA is cut into pieces at these sites. A bacterium protects its own DNA from being attacked by adding methyl (CH_3) groups to those sites.[10] Arber recognized the potential significance by suggesting that a restriction enzyme might "provide a tool for the sequence-specific cleavage of DNA." Botstein realized that the distinctive patterns of fragments produced by these enzymes could be used for mapping the genome.

In one of those ironic twists, the enzymes of the original bacterial system did not prove useful for DNA mapping, but other restriction enzymes discovered in 1970 by Ham Smith at Johns Hopkins University did.[11] A year later Dan Nathans, also at Johns Hopkins, showed that Ham Smith's enzyme could be used to cleave the monkey virus SV40 into a set of fragments that could be distinguished using a technique called gel electrophoresis (in which fragments are separated by their size).[12] Arber, Smith, and Nathans shared a Nobel Prize in 1978.

The latest database of restriction enzymes has more than 50,000 enzymes, with more than 600 different sequences identified as targets.[13] Target sequences are short, usually between 4 and 6 bases long. The crucial insight in the Botstein proposal was that a mutation in one of these sites must change the pattern of cleavage produced by the enzyme. The changes are called RFLPs (restriction fragment length polymorphisms), and they can be treated in the same way as any mutation with a visible effect.[14]

The problem with a linkage map is that it puts the target loci in order, whether they are mutants or RFLPs, but doesn't represent actual distance along the chromosome. All sorts of distortions affect the relationship between recombination frequencies and real distance along DNA. (The length of DNA is measured in base pairs; kb stands for kilobases [1000 bp], Mb measures in millions, and Gb measures in billions.)

To sequence the genome, we need a *physical map*, based on actual distance along DNA. This can be achieved using the fragments of DNA produced by restriction enzymes. The process begins by looking for overlaps between the fragments produced by different restriction enzymes. This enables the fragments to be placed in an order. By lining up the fragments, it's possible to measure actual distances.

The technique of DNA fingerprinting uses the same principle as restriction mapping in separating fragments of DNA by size. Fragments are generated by restriction enzymes, but instead of coming from mutations at the target sites, the variation is caused by differences in very short repeated sequences called minisatellites that lie within the fragments. Minisatellites change frequently, and their pattern is different in every individual, so it can be used to identify children and parents, or for forensic purposes to identify a person who left DNA at a crime scene.

Here is another example of the Law of Unintended Consequences: Alec Jeffreys was trying to find the cause of the extreme variation at these sites in 1984. He recalls that his first thought was, "[That's] a horrible, smudgy, blurry mess." Then he realized that the bands are unique identifiers. Capturing the way science works, he said much later, "If someone had told me in 1980, 'Alec, go away and figure out a way of identifying people with DNA,' I would have sat there looking very stupid and got nowhere at all."[15]

Alec Jeffreys (now Sir Alec Jeffreys) came into the field of mapping with restriction enzymes when he was mapping the ß-globin gene at the University of Amsterdam. This was just at the time when split genes were discovered. When he returned to England in 1977, to the University of Leicester, he decided that he could not compete in the gene mapping field—there was a stampede of much larger laboratories into the field from all over the world—and he started to look for variations in restriction fragments between individuals. That led to the discovery of minisatellites.

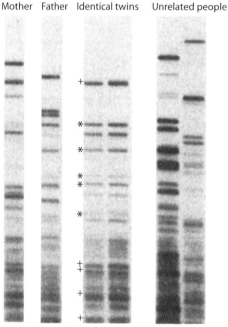

Mother Father Identical twins   Unrelated people

* = bands in twins inherited from father
+ = bands in twins inherited from mother

An example of DNA fingerprinting from the original research paper. Every individual has a different pattern, except for identical twins. The twins have some bands inherited from their father, some inherited from their mother, and some new bands. There is no relationship with the pattern from unrelated people.[16]

The first practical use followed rapidly. In 1985, an immigration lawyer contacted Jeffreys to ask if the technique could be used to confirm the identity of a boy who had been denied entry to the United Kingdom although his (supposed) mother lived there. The test proved the point and the boy was allowed in. Forensic use followed in 1986, when the police asked Jeffreys to use the technique to prove the identity of a supposed rapist. The test showed he was innocent of both rapes. Now, of course, we live in a brave new world where it would in principle be possible to establish a database to identify every living individual by their DNA.

A physical map of DNA needs to be related to the actual structure of the chromosomes. Until the development of *banding techniques* in 1971, chromosomes were essentially distinguished only by their size, as seen during cell division. It's sometimes hard to realize how primitive our knowledge was then. When it was discovered that treatments with various agents can produce a pattern of "bands" in the chromosome, it became possible not only to identify each chromosome unequivocally, but also to identify cases in which material had been translocated from one chromosome to another. Specific DNA sequences, whether they represent known genes or RFLPs, can be mapped to the bands, providing a crude connection between the linkage map and the physical map.

The first large-scale approaches to genomes were contemplated in the early 1980s. They were somewhat driven by objectives of pushing the limits of existing technology[17] or creating huge facilities. Robert Sinsheimer at the University of California, Santa Cruz held a meeting in 1985 to consider the possibilities for the human genome, with the idea of creating a facility at Santa Cruz to do the work. Sinsheimer had purified the bacterial virus φX174, which had been the first DNA to be sequenced, and there were several large-scale science projects at Santa Cruz.

Right from the start, the idea was an overt proposal for "big science." "Biology had always been 'small science.' I wondered if there were scientific opportunities in biology that were being overlooked, simply because we were not thinking on an adequate scale," Sinsheimer said.[18] No one at the time could have envisaged the full scale the project was ultimately to achieve.

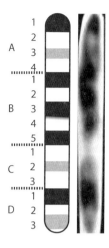

The most versatile chromosome banding technique is Giemsa staining, named for the German biochemist Gustav Giemsa, who developed the stain in 1904. The bands are called G-bands. Mouse chromosome 10 has 15 G-bands (numbered and pictured on the left, an original stained example on the right). There are approximately 300 bands in the entire set of mouse chromosomes. The average band has ~10 Mb.[19,20]

"The total human sequence is the grail of human genetics," Wally Gilbert said.[21] One of the participants in the Santa Cruz meeting, he had shared a Nobel Prize with Fred Sanger in 1980 for their work on DNA sequencing. His credentials in biotechnology dated from his role as a cofounder of Biogen, one of the earliest biotech companies, in 1978. In a theme that was to be repeated some years later, impatient with the pace of progress under the public aegis, he subsequently proposed to establish an independent company to sequence the genome. His attempt to establish the Genome Corporation in 1987 was not successful, however, and he returned to advocating the public effort.

The path to the Human Genome Project after it was proposed in 1985 was neither smooth nor predictable.[22] As the proposal reached a wider audience, it became controversial. When it was discussed in an informal session at the 1986 Cold Spring Harbor Symposium (the subject of the Symposium was the Molecular Biology of *Homo sapiens*), there were concerns that the project might compete for resources rather than complementing existing efforts, and that the introduction of "big science" could change the nature of "small science."

As David Botstein from MIT said, "If it means changing the structure of science in such a way as to indenture all of us, especially the young people, to this enormous thing like the space shuttle, instead of what you feel like doing, or even like they used to do in little companies—a third of your time you sequence and the other two-thirds of the

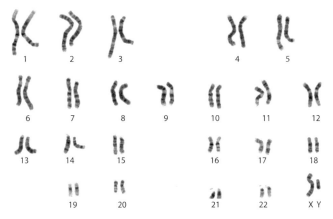

The human karyotype (the entire set of chromosomes) has 22 pairs of autosomes (nonsex chromosomes) and a sex pair (XX in female, XY in male; the Y is much smaller than the X). Chromosomes are shown during mitosis (cell division) and are stained by the Giemsa technique that distinguishes each chromosome by a distinct set of "bands."

time you can do whatever you want, okay? When you think about it in those terms, the question is: What's the price? What's the price? And I think that that's really the relevant question."[23]

As for the nature of the project, there was a view that sequencing the entire genome might be using a sledgehammer to crack a nut. "It seems to me that not all sequences are equally valuable," David Botstein said. Wally Gilbert thought the project should start with the most interesting sequences. "I would subdivide into the first one percent of sequence, which is thirty megabases. And this is probably all genes, all DNA. And the second ten percent, which is three hundred megabases.... In fact, this first one percent is probably all of the most interesting biological properties.... The next ten percent and the next ninety percent is unfortunately most of the physical problems." Some people felt that, given this view, it might be better to concentrate on making a physical map, and then to sequence the important parts.

Projections for how long the project might take tended to be pessimistic. Wally Gilbert estimated that "the rate of DNA sequencing today, let's say around the world in the entire scientific community, is of the order of 2 Mb/year. So it would take something of the order of 1,000, 1,500 years to sequence this amount of DNA.... With a gradual improvement of techniques... it will take probably of the order of 100 years." Dave Smith of the DOE (Department of Energy) felt that it was "not unlikely that within a few years we can increase this sequencing rate by a factor of one hundred. And that puts this idea in an entirely different framework."

The DOE was involved because it made the first proposals for a massive public sequencing effort. This was in itself a statement that the nature of the project would be quite different from anything that had preceded it in biology. "The Department of Energy has successfully managed many long-term and complex technological programs.... The size, interdisciplinary nature and long-term scale of the Human Genome Project, with the many technologies involved, fits these experiences of DOE well."[24] (The DOE's interest in a biology project was triggered by the responsibility it had been given for assessing the dangers of radiation after the atom bombs were dropped in 1945.)

There was some relief at the Cold Spring Harbor meeting at the prospect that the DOE would manage the project rather than the NIH. (The National Institutes of Health is the major source for funds for biomedical research in the United States.) As Wally Gilbert said, "We do not want to have the NIH to be in our debt running such a project, because in that case it is in direct competition with the RO1 money [Research Project Grant program at NIH]. Whereas if it's being run by DOE, if in fact their separate agencies of the government can compete, you can at least separate the project from your other sources of funding."

As the project gathered steam, conflict broke out as to whether DOE or NIH should spearhead the effort. The first funds were appropriated for the project in 1988; of the $17 million, $12 million came from new funds and $5 million was diverted from existing programs.[25] Over the first five years, the genome project brought in about $400 million of new funding and the NIGMS (genetics) program at NIH lost about $25 million.[26] By the time the project formally began in 1990, at several dedicated genome centers, it had a budget of $87 million, two-thirds controlled by NIH, almost all of which came from new appropriations. There was a funding crunch at the time, partly caused by the large proportion of NIH funds spent on AIDS, partly by administrative changes in procedures, but opinion moved slowly in favor of the project.

Genome sequencing centers were set up (in Europe as well as the United States), and there was open warfare as to what this meant for the biology community. Extreme positions were epitomized by an exchange at a meeting on the genome in 1990 when Don Brown of the Carnegie Institution said that the project was "overtargeted, overbudgeted, over-prioritized, overadministered, and has to be micromanaged [whereas] the [RO1] investigator-initiated grant system at NIH … has been the absolute pride of the biomedical enterprise." Jim Watson retorted that, "most of [the RO1s] aren't that great anyway."[27]

The extremes of positions are captured by Tom Caskey's recollection of the fate of a sequencing proposal. "In 1986, we were making very good progress with semi-automated cloning and sequencing. Craig Venter and I had put forward a proposal to sequence the X chromosome and identify disease genes. We had a reverse site visit (when the investigators go before the grant committee). When it was our turn, the Chairman of the committee said, 'We won't be needing to hear the next proposal because it's based on automated sequencing, and we don't know if that will work.' When I told Jim [Watson] about this, he got very agitated and said, 'but the genome project is dead if we don't have auto-mated sequencing.' I told him not to be too concerned, but not to let the Chairman sit on any of his grant committees."[28]

"The paradigm is shifting," Wally Gilbert said. "Twenty years ago, every grad student working on DNA had to learn to purify restriction enzymes. By 1976 no grad student knew how to purify restriction enzymes, they purchased them…. Science will not be less experimental, but it will be different experiments…. Classic biochemistry … will no longer exist." He thought the genome project would mean that research-ers would no longer isolate a gene and sequence it in order to do exper-iments; instead they would look up the information in a database in order to design experiments. Critics were confusing science with tools, he concluded.

This is the heart of the matter. Was it a clash between Luddites, wedded to traditional ways of performing science in biology for their own sake, standing in the way of progressives, who wanted to move to more productive, modern methods? Or was it a clash between those who wanted to preserve an intellectual approach that had yielded major insights (including many of direct relevance to human welfare) and those who wanted to move to mass-production methods out of a love for modern technology? Jim Watson had the last word when he said, "You have to realize we are talking religion."[29]

CHAPTER

17

GENOMES

I would only once have the opportunity to let my scientific life encompass a path from double helix to the three billion steps of the human genome.[1]

Jim Watson, 1989

This chapter is called *Genomes* and not *The Human Genome* because one of the more surprising results of the Human Genome Project was the extent to which every human genome is different. Of course, we knew that every individual (except for identical twins) has a different set of genes, but we had no idea at the start of the project that there would be such extensive differences in other parts of the genome (also revealed by DNA fingerprinting as discussed in the previous chapter). Indeed, we had no idea that genes themselves would comprise such a small part of the genome.

The state of play at the start of the project was that only a limited number of human genes were known; in the edition of his handbook of human genes published in 1990, Victor McKusick estimated this was "perhaps only 5% -10% or less of the structural genes."[2] Humans are more complicated than other organisms, so it seemed obvious they should contain the most genes. When the project was proposed for sequencing the human genome, it was thought that 100,000 genes would be found.[3] It was a surprise (and perhaps a disappointment?) that we have barely more than a worm, the same as the mouse, and fewer than a mustard plant.

Looking back, you can see how the assumptions of the era were misleading. It was thought that most of the genome would comprise genes coding for proteins, and that most of the RNAs would be intermediates in the production of those proteins. Neither was correct. In fact, there are about an equal number of sequences coding for RNAs as for proteins

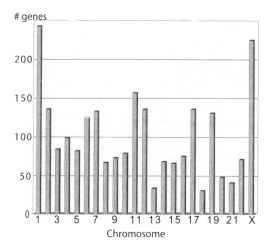

There were 2300 known human genes in 1991, distributed unevenly among the chromosomes. No genes were known on the Y chromosome at the time.[4]

(although in most cases the RNAs have no known function). It took analysis of the actual sequence to come to grips with the issue. Less than 1% of the human genome codes for protein, and <25% is implicated directly in gene expression. There has been lots of speculation as to why there is so much DNA in the genome, but the straight answer is that we still don't know.

Jim Watson became the first Director of the Human Genome Project in 1988, which then powered forward, with a certain sense that this was the wave of the future. "Most knowledgeable people and most eminent scientists are solidly behind the genome project," said James Wyngaarden, who was Director at the NIH. "The ones who are critical are journeymen biochemists who may be having a hard time competing themselves."[5]

Having Jim Watson as Director gave the project a terrific impetus. He had turned Cold Spring Harbor from a minor laboratory into a major institution. When I first went there, to the Symposium in 1970, to say it was shabby would be kind. Today it has more the feeling of a wealthy upscale liberal arts college in New England.

Jim is nothing if not forthright. I remember a conversation, I think it must have been in the early 1990s, when he came up to me at a meeting at Cold Spring Harbor and said without any preamble, "You're my enemy." I was too taken aback to make much of a coherent response, but it emerged he was cross about a dispute that had arisen between *Cell* and

Cold Spring Harbor about double publication of data in the journal and the Symposium. "You're affecting my livelihood," he said. Perhaps nothing but a full throttle approach would have jammed the Human Genome Project through. (However, there was a sad end to Watson's career in 2007 when he resigned from his position at Cold Spring Harbor after expressing the view that Blacks have lower intelligence based on genetics.)

The project became an international effort, largely powered by the United States, with Britain in a substantial second place, Europe somewhere behind, and the rest of the world playing minor roles. HUGO (Human Genome Organization) was established as a coordinating body, running workshops in which progress could be assessed. This was top-down directed management, with each center working on assigned parts of the genome (or other genomes as the project broadened to include other species). This form of organization was new to biology, although it would have been familiar, for example, to physicists.

The pace of sequencing increased (and the cost per base pair decreased) in an astonishing way during the next decade. Sequencing had started with proteins in a distinctly artisanal way in 1945, when Fred Sanger, working in Cambridge (England) started to analyze insulin, with a painstaking approach requiring analysis fragment by fragment. It was not certain at the time that a protein had a unique sequence, but 10 years later, the sequence of the two insulin chains was complete, and the point was proven.[6] Using the Edman method, which removes one amino acid at a time from the end of the protein, automated protein sequencers were developed in the 1970s.[7] Known for his retiring nature and modesty, Sanger once said of himself, "Unlike most of my scientific colleagues, I was not academically brilliant."[8]

First attempts at nucleic acid sequencing applied comparable methods to small RNAs. But applying these methods to DNA was impractical because of its large size. Sanger turned to another method in 1973. DNA was copied (using the enzyme DNA polymerase) to make new strands that stopped at a specific nucleotide. They all started at the same point, so the length of each chain identifies the positions of the target nucleotide. Doing this four times, once for each nucleotide, identified the positions of each of the four bases. Fragments are separated by their length on a gel electrophoresis column. Wally Gilbert's laboratory at Harvard developed another method, in which the DNA chain is broken at specific bases.

The length of DNA that could be analyzed on one gel increased, and the technology moved from visual reading of the gels to automation. The first automated DNA sequencers came on the market in 1986. Originally, the chains were identified by radioactive labeling; radioactivity was replaced by fluorescent dyes, and gels were replaced by capillaries, when the automated methods of

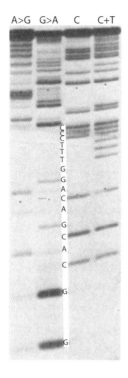

The four lanes of a Maxam–Gilbert sequencing gel show A > G (both A and G react but A is stronger), G > A (G reacts more strongly than A), C (only C), and C + T (both C and T). The sequence can be read as shown in the center.[9]

"next-generation sequencing" were developed. Instead of multiple lanes, all the bases can be read in a single lane. Today's machines can read kilobase sequences in one reaction.

Sequencing is now into third-generation techniques, using nanopore methods. The record for the longest sequence achieved in a single read (known as the read-length in the trade) is now 2.3 Mb.[10] The cost is way below the original target of $1 per base, and the speed has reached a point at which one recent research paper reported the results of sequencing 10,000 individual human genomes.[11]

Sequencing has become a massive effort. In their latest iteration, sequencing machines are available as a series from those suitable for small-scale efforts to those that can undertake multiple genomes at a stroke. The latest development is the ability to distinguish methylcytosine as well as the standard four bases, so the methylation pattern of a whole genome can be readily analyzed. (Methylation of cytosine is a modification of DNA that has important implications for controlling gene activity, as discussed in Chapter 19.) RNA-seq is an abbreviation for RNA sequencing,

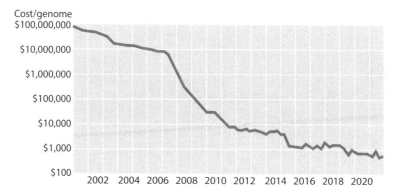

The cost of sequencing a human genome has decreased from $1 billion in 2000 to less than $1000 today.

a modification that allows all the RNAs of a cell to be sequenced. You no longer look at data when you do sequencing, just read the printout.

Just as the ultimate outcome of research cannot be projected, technological advances can have unanticipated spin-offs. Sequencing was seen as a means to assemble a basic set of data that would drive both research and medical advances. Although it wasn't predictable that the cost of sequencing would fall so low, the development of services such as 23andMe for genetic testing by DNA sequencing seems a natural outcome. On the other hand, no one foresaw the use of sequencing as an essential tool in handling the COVID pandemic in 2021, yet without rapid sequencing, diagnosis and the response to new variants would have been vastly more difficult.

Sequencing is half the battle in defining the genome, perhaps the last half. The first half has been obtaining a set of ordered fragments from the genome that can be sequenced. (Of course, this becomes easier as the read-length becomes longer.) Here the transition from hypothesis-driven science to science driven by technology (in this case the methodology of DNA sequencing) was accentuated by a debate about the best way to do this.

The first plan to sequence the human genome devoted the first five years, 1991–1995, to improving and developing the technology, with the intention of completing the sequence over the following 10 years. At $200 million per year, the total cost would be $3 billion. (This was unprecedented in biology but small by the standards of "big science" in physics. The cost of building and running the CERN particle accelerator to find the Higgs boson in 2012 was about $13.25 billion.)

The approach to sequencing the human genome was conventional, with the genome first broken into ordered fragments, followed by sequencing of the fragments. Craig Venter was one of the leaders in sequencing at the NIH, but left over disagreements about the technology. He founded an independent institute nearby and turned to an approach called shotgun sequencing, in which the genome is fragmented randomly into fragments. Each fragment is sequenced from both ends. By using *oversampling*, in which in effect the total length of DNA sequenced is several times the length of the target genome, software can be used to assemble a unique sequence from the overlaps between fragments.[12]

Venter applied the technique successfully to the 1.8 million base pair genome of the bacterium *Haemophilus influenzae*.[13] In 1999 Venter started on the fruit fly genome. "It took only four months to sequence the genome instead of 13 years for *E. coli*, 10 years for yeast... . Key to that (were) the new algorithms that made it possible to assemble the DNA," Venter said.[14] (This was, however, misleading: in fact, it took four months to obtain enough samples to cover the whole genome of the fly, and the completed sequence was published a year later.)

Shotgun cloning was controversial to the extent that Venter's grant proposal had been rejected. (Some people thought it would not

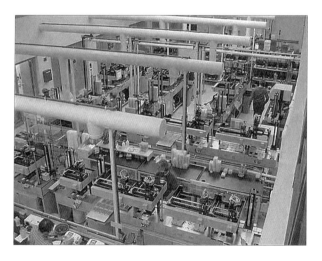

An automated DNA-sequencing center.

be possible to find all the overlaps to join the sequences together.) It was even more controversial when Venter proposed to apply the same approach to the human genome. As the Genome Project advanced, plans were revised, and in 1998 a final five-year plan was made, intending to sequence worm and fruit fly, and one-third of the human genome by 2001, with completion by 2003. This relied on the conventional approach of ordering the fragments before sequencing.

Venter decided he could complete the project more quickly, at much less expense, with shotgun cloning, and founded a company, Celera, to undertake the project. Not only was the scientific basis controversial, but the concept that a private company would profit from the genome sequence was anathema to many scientists. And even more threatening, the thought that a private company could do this led to questions as to whether public money should continue to be spent.

Venter offended virtually everyone involved with the Human Genome Project when he held a press conference and suggested that the Human Genome Project should leave sequencing the human genome to Celera and instead focus on sequencing the mouse genome. So Celera sequenced the genome in competition with the Human Genome Project.

Celera used 300 automated DNA sequencers to generate 14.8 Gb of sequence (the machines came from Applied Biosystems, which owned Celera). This represented fivefold oversampling of the 3-Gb human genome sequence. Celera combined their data with the earlier results of the Human Genome Project to generate a consensus sequence covering 94% of the genome.[15] It was published simultaneously with the paper from the Human Genome Project in 2001.[16]

These papers marked a reversal of the tradition that the (junior) scientist who did most of the work is the first author and that the senior author comes last: Craig Venter was first author on the Celera paper and Eric Lander (a leader of the U.S. sequencing effort) was first author on the page listing the authors of the International Human Genome Sequencing Consortium. There is no individual researcher to acknowledge for the brunt of the work, and the paper becomes identified with the principal organizer. Does this mark a turning point for biology's transition to big science?

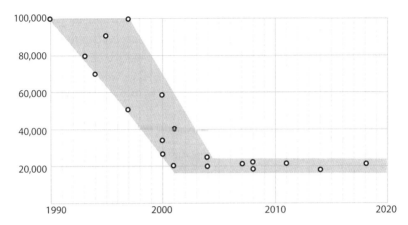

Estimates for the number of human protein-coding genes dropped steadily from 100,000 in 1990 to the end of the century, and since then have oscillated around 20,000.[17,18,19]

These first human genome sequences were called draft sequences, because they were not quite complete. Each estimated the number of protein-coding genes at just under 30,000, but subsequent comparisons showed that the novel genes (those identified solely on the basis of sequence) were somewhat different in the two sequence sets.

This makes another point: it's not the raw sequence that's the end point but the *annotation* that is critical. Annotation is the vernacular of the trade for marking up the sequence to identify genes and other features of interest. This requires a completely different skill set from sequencing itself.

The objective of the Human Genome Project was to obtain a sequence with 99.99% accuracy: the competition forced a compromise to a lower standard, publishing the draft sequence rather than a complete sequence. Competition from Celera also forced the Human Genome Project to advance its completion date by several years. Politics instead of intellectual criteria determined when to publish.

John Sulston, who was in charge of the British sequencing effort, describes the difference in the atmosphere from conventional scientific research. "The sequencers were no longer running traditionally structured labs, with a group of more or less independent scientists and a few technicians in support: we were effectively running 'businesses'... . It was not just a matter of being on your own and so able to chuck everything out of the freezer and start again if a line of research didn't work out—which is the way you should do science."[20] Responsibility for a trained staff of hundreds of people and commitment to a major funding program created a different set of imperatives.

It's difficult to assess the results critically, because the number of genes depends largely on annotation, which comes from software-driven analysis of the data. The estimate changes not only when the sequence is revised, but also when there are improvements in the software. No impartial third party without the ability to analyze and modify the software can really assess how robust the number is. This is not directly in conflict with the principle that a scientific paper should contain sufficient information for others to repeat and verify it, but it certainly makes it more difficult.

What does this mean for the self-correcting mechanism of science? The number of human genes has been revised at least nine times since the year 2000. Different research groups use different criteria. Eventually, a consensus will emerge but possibly without ever defining what computational factors or assumptions led to erroneous estimates. Self-correction still works, but less directly.

Even critics of the "big science" approach of the Human Genome Project agree that the sequences of the human and other genomes are a fantastic tool for further research. Analysis of the genome itself makes it easier to devise diagnostic tests for disease genes, although it does not of itself lead to cures for inherited or other diseases. And in addition to their intrinsic interest, comparisons between genomes of different species throw a crucial light on evolution.

Genome sequencing places previous ideas about the relationship between humans and our nearest relatives (chimpanzees) on a firmer basis. Comparing protein sequences suggested as long ago as 1975 that differences between proteins are minor: many proteins are identical, and the average human protein differs by only one amino acid from its chimpanzee counterpart.[21] This makes it seem that the answer to the difference between human and chimpanzee does not lie in protein sequences.

Comparing the genome sequences frames the difference in a more quantitative way. Human and chimpanzee genomes are very

■ SEQUENCES OF GENOMES OF SOME IMPORTANT SPECIES ■

SPECIES	GENOME	PREDICTED GENES	YEAR	AUTHORS
Escherichia coli (bacterium)	4.6 Mb	4288	1997[23]	18
Saccharomyces cerevisiae (baker's yeast)	12.1 Mb	6294	1996[24]	16
Caenorhabditis elegans (worm)	100 Mb	19,735	1998[25]	147[26]
Drosophila melanogaster (fruit fly)	165 Mb	13,600	2000[27]	197
Arabidopsis thaliana (thale cress)	13 Mb	25,498	2000[28]	128
Homo sapiens (human)	3.2 Gb	19,042	2001[29,30]	275
Mus musculus (mouse)	2.9 Gb	20,210	2002[31,32]	147

closely related, often stated to be 99% identical, which is almost right. Comparing the corresponding sequences, the variation is 1.2%. In addition, there have been rearrangements, insertions, and deletions, amounting to another 1.5% of differences.[22] This may not sound a lot, but still amounts to more than 30 million individual changes. (The most dramatic rearrangement is the fusion of chimpanzee chromosomes 2A and 2B into the single human chromosome 2, explaining why chimpanzees have 48 chromosomes and humans have 46.)

Genes expressed in neurons, especially those that control gene expression, show greater variation than average between human and chimpanzee. There is a greater rate of substitution at CpG islands, a feature of the genome sequence potentially associated with control of gene expression, as discussed in Chapter 19. Perhaps humans and chimpanzees have essentially the same genes but different methods of controlling them.

We enter another world in defining the relationship between *Homo sapiens* and Neanderthal man. This is uniquely possible because of genome sequencing. By extracting DNA from Neanderthal bones, Svante Pääbo showed that *H. sapiens* was very closely related to Neanderthals; the genome sequences are only 2% apart on average.[33] (Pääbo was awarded a Nobel Prize in 2022.) Subsequent work went on to show something that could never have been deduced from the bones: there is an average 2.5% of Neanderthal DNA in the human genome. (Some humans have more Neanderthal DNA than others.[34]) This is probably because 50,000–60,000 years ago, humans and Neanderthals were still interbreeding.[35] Comparisons of genome sequences between present-day humans, Neanderthals, and Denisovans (the first extinct human group defined solely on the base of DNA sequences) offers the potential for using triangulation to gain new insights into humanity.[36]

"[It] is beginning to become feasible ... to exploit the fact that Neanderthals and Denisovans have contributed variants to the present-day [human] gene pool," Pääbo says.[37] Sometimes those genes are ancestral versions, distinguished from the modern versions that almost all humans carry. "The problem is that those [ancestral] contributions are often of low frequency so that one needs very big cohorts ... to be able to see an effect."

This is where "big science" can offer insights that would never have been possible from conventional research. One example is the identification of a gene in which the Neanderthal version increases sensitivity to peripheral pain; it's present in 0.4% of modern Britons. This illustrates the interaction of big data with conventional research, as the identification of the gene from the sequence data was followed by an analysis of the electrophysiological effects of the protein it codes for.[38]

Does the widening gap between original data and interpretation in "big science" threaten the integrity of science? The increasing complexity of research depending on collaborations between researchers with different areas of expertise makes it hard to get an overview for either researchers or reviewers. But the principle that science will correct itself remains valid, even if this is no longer accomplished by direct attempts to repeat the same experiment. Take the example of the two versions of the human sequence produced by the Human Genome Project and Celera. Differences between them led to further papers and eventually there will be a consensus. Of course there is no unique human genome for the species: every individual is different. The nature of those differences is one of the fascinating features to come out of the capacity to sequence many individual genomes.

CHAPTER
18

EDITING

I see all developments that have emerged as another proof of how unpredictable the benefits derived from the genuine search for knowledge are.[1]

Francisco Mojica, 2021

CRISPR is a striking demonstration that you never know where science will lead. It was discovered as a system that gives bacteria immunity against viral infection. Viewed as rather arcane, the early papers were all published in somewhat obscure journals. Like the systems of restriction enzymes that bacteria use to protect themselves against viral infection, discussed in Chapter 16, which were discovered by researchers interested in viral–bacterial infections, it transmogrified into a major system that can be adapted to work with human material. It provides the most powerful method yet discovered for gene editing— changing genetic information. Indeed, this seemingly irrelevant piece of research led ultimately to a fight between several groups to obtain patents for using it and debates about its potential application to human gene therapy. What price arcane research?

It's fascinating to trace the path from an observation that was ignored, to the realization that it might be involved in interactions between a virus and bacterium, to the possibility that this might be a new type of enzyme system with practical implications, to disputes about priority. It first attracted wider interest because of the mistaken belief that it might be a new type of system using a phenomenon called RNA interference (when small RNAs are involved in controlling gene activity). As soon as it was realized that in fact it works on DNA, there was a stampede into working on the system. It's a demonstration of the gravitational pull of new ideas in science.

233

```
GTTACAGACG AACCCTAGTT GGGTTGAAGC GAACAGGATG GCGAACCGGT GTCTGCCACCA GTT
GTTACAGACG AACCCTAGTT GGGTTGAAGC CACGACAATC AAGTCTGGTT GCATGGCGAC ACGGA
GTTACAGACG AACCCTAGTT GGGTTGAAGC CTGTGCCTCC AGCGGCCGTC AGACAGTCGC ATCCGA
GTTACAGACG AACCCTAGTT GGGTTGAAGC AAGAAGCCGC TCGCCGTCCT CGATGACGGG CGGGCG
GTTACAGACG AACCCTAGTT GGGTTGAAGC GACAAGACTC GCGACGAAGC CGAGTCGAAA CGCCGC
GTTACAGACG AACCCTAGTT GGGTTGAAGC CTCTTTATCC CTCCTGCCCG AATGTCTACG AATATC
GTTACAGACG AACCCTAGTT GGGTTGAAGC GAACCCACTG GTGAAGAAAA AGTTGTAGAG ACCCTA
GTTACAGACG AATCCCTAGTTGGGTTGAAGC ACGACAATCA AGTCTGGTTA CATGGCGACA GGATGG
GTTACAGACG AACCCTAGTT GGGTTGAAGC TTCCACAACG TCGGGGAGGG CGAAATTAGC CAAGCA
GTTACAGACG AACCCTAGTT GGGTTGAAGC TCCCGCTGGG GATGTCGGGA GTGCCGGGCG AGCCA
GTTAGAGACG AACCCTAGTT GGGTTGAAGC CCCGGCCCGT TGCCCCCCAC GGCAATCGTC TGCT
GTTACAGACG AACCCTAGTT GGGTTGAAGC GGTCTGTGTT ATTCTGTGCG TCTGCCGCGA CAAC
GTTACAGACG AACCCTAGTT GGGTTGAAGC ATTGCCTGTA CCCGTCGTGT AATCTTAGTCC GAATG
GTTACAGACG AACCCTAGTT GGGTTGAAGC GAGATGTGCG ACCGCGGCGA AATGAGCAGT TCGTG
GTTACAGACG AACCCTAGTT GGGTTGAAGC GCGACATGGG GACCGTCGAG AACGCGCTCT ATGGGGA
GTTACAGACG AACCCTAGTT GGGTTGAAGC CGAGGGTCCC GGTGTCGAGA GGACCGGGAC GGACGGA
GTTACAGTCG AACCCTAGTT GGGTTGAAGC TCGGTAATCT GGGAAGGCGT CAGTCTCGGC CGAGTAATC
GTTACAGACG AACCCTAGTT GGGTTGAAGC CTCGCCATCG CCGCGAACTC GGTCCTCCTC GGGGTG
GTTACAGACG AACCCTAGTT GGGTTGAAGC AAGCCTTGAG AGTGTCTGTT GGTATGATGA ATGTT
GTTACAGACG AACCCTAGTT GGGTTGAAGC AAGTAGACCG CGCTCAGTTA CGACAGCTGC TCGA
GTTACAGACG AACCCTAGTT GGGTTGAAGC ACGATGATCT CGCCAGTCTG CAGCGTTACA TTGG
```

30 bp repeats   unique spacers

Tandem repeats in archaea were discovered in 1995. The continuous sequence is shown as a series of rows for clarity. Each boldface sequence is one repeat. Each copy of the 30-bp repeated sequence is separated from the next by a spacer of 33–39 bp. The sequence is broken into groups of 10 bases to make it easier to read.[3]

The observations that led to CRISPR had roots two decades earlier. Working as a PhD candidate at the University of Alicante in Spain in 1995, Francisco Mojica found a series of tandem repeats in the genome of a prokaryote (prokaryotes are organisms, such as bacteria, that do not have a nucleus). Long stretches of DNA were occupied by many copies of the same sequence of 30 bp (base pairs), repeated at regular intervals. Mojica found that Yoshimuzi Ishino had observed a similar repeating structure in a gene of the bacterium *Escherichia coli* in Japan in 1987.[2] Ishino had commented that they had no known biological function, and Mojica's proposals for function were wonderfully wrong.

The system was regarded as so arcane that Mojica had difficulty getting funding to continue work on it, but by 2000 he was able to demonstrate that repeats of this type were common in prokaryotes.[4] He called them short regularly spaced repeats (SRSRs). The actual sequences were quite different in different organisms, but the pattern of repeats of 21–37 bp separated by unique sequences of similar length was common. They were renamed as clustered regularly interspaced short palindromic repeats (CRISPR) when Ruud Jansen in Utrecht showed that they were flanked by a set of four conserved genes in a variety of prokaryotes. The genes were called *cas1–4* (CRISPR-associated genes).[5]

The first indication of a function for the system, in fact the proposal that it was part of a bacterial immune system, came when Mojica discovered that the spacers between the CRISPR repeats were present in the sequences of bacterial viruses.[6] The idea was not well-received. Top journals rejected the paper. *Nature* rejected the paper without review on the basis that the idea was already known; the *Proceedings of the National Academy of Sciences* said it lacked sufficient novelty and importance.

Moving down the hierarchy of scientific journals, the paper was in turn rejected by *Molecular Microbiology* and *Nucleic Acid Research*. Finally it was published in the *Journal of Molecular Evolution*. Failure to appreciate its significance was not an isolated incidence, because a paper from a French group making a similar proposal[7] was also rejected from a series of journals.[8]

What were the editors of the journals thinking? The papers reported a provocative correlation, but did not prove the existence of a novel immune mechanism. There was reasonable evidence to suppose that the spacer sequences originated in external elements that infect bacteria. Among the examples were some in which the bacteria were not infected by viruses that had the sequences of the CRISPR spacers. Possibly there may have been a view, looking for "relevance," that bacterial immune systems are no longer interesting.

It would have required an unusually perspicacious editor or reviewer to draw a parallel with restriction enzymes and argue that a bacterial immune system might have relevance to humans, and if this had been suggested in the papers, it would have been laughed out of court: hindsight is easy. As science has moved deeper into an era in which papers are expected to present massive amounts of data, the criteria for presenting novel ideas have become progressively steeper. In any case, it wouldn't have been unreasonable for general interest journals to require more conclusive evidence of mechanism, but it's surprising there was such a problem with journals lower in the hierarchy, posing the question of whether there might be some intrinsic bias against lesser-known laboratories.

Yogurt provided the key to understanding the function of the system. Lactic acid bacteria are important in the production of dairy products, including yogurt and cheese, and foodmakers were concerned about viral infections that destroyed the bacterial cultures. A team led by Philippe Horvath at Danisco in France saw a correlation between CRISPR spacers and resistance to viruses in *Streptococcus thermophilus,* one of the bacteria used in producing yogurt and cheese. They then demonstrated directly in 2007 that bacteria acquiring resistance to a virus have new spacers whose sequences were present in the virus

"These findings suggest that the presence of a CRISPR spacer identical to the phage sequence provides resistance against phages containing this particular sequence," they concluded. They also showed that resistance was lost when they inactivated certain *cas* genes, suggesting that the spacer sequence provides the specificity, and *cas* genes code for the enzymatic apparatus.[9] This paper was published in a general interest journal, *Science,*[10] and CRISPR came into general view. From now on, most of the significant papers were published in high-profile journals.

The significant actors in the next stage of development were attracted to the system by the thought that it might operate by RNA interference. As it turned out, this is not the way CRISPR works.

The potential for gene editing was realized as soon as the basis for specificity was established. Working in Erik Sontheimer's laboratory at Northwestern University, which had been involved in RNA interference, postdoc Luciano Marraffini performed an ingenious experiment in 2008, suggesting that the target for the system was in fact DNA, not RNA.[11] This meant that in effect, the system functions as a restriction enzyme that is programmed by the spacer sequence. If you wanted to draw an analogy with artificial intelligence (AI)-driven software, you might say that the system has pattern recognition trained by patterns that it has previously encountered.

Sontheimer and Marraffini applied for a patent. It was rejected, because it was an idea, without any proof of concept.[12] A grant application to the National Institutes of Health to investigate the potential use for gene editing was also rejected. The full practical and commercial implications became accepted only later.

The attack on the system had been mostly genetic, and further progress required a move to biochemistry. The demonstration of sequence specificity in the system brought in new players.

A team led by John van der Oost in the Netherlands produced the first artificial CRISPRs in 2008 by incorporating sequences corresponding to the bacteriophage (virus) lambda into the array. Bacteria with the artificial CRISPR became resistant to lambda.[13] It was a sign of how far, and how quickly, things had advanced that their paper started with the simple statement, "Prokaryotes acquire virus resistance by integrating short fragments of viral nucleic acid into clusters of regularly interspaced short palindromic repeats (CRISPRs)."

RNA does play a key role in the specificity of the system. The van der Oost team showed that an RNA, called pre-crRNA, is transcribed from the repeat-spacer array at the CRISPR locus, and that one of the *cas* gene products cleaves the longer RNA into smaller crRNAs of ~57 bases.[14]

Horvath's group in France (now at DuPont Nutrition, which had bought Danisco) teamed up with Virginijus Šikšnys at Vilnius University in Lithuania. By expressing the *S. thermophilus* system in a different bacterium (*Escherichia coli*), they were able to show that the Cas9 protein plus crRNA can cleave a target DNA that has a copy of the spacer sequence represented in the crRNA. The crRNA works by binding to a sequence that matches it in the target DNA. *Changing* the spacer sequence directs the system to a new target, one that matches the new spacer sequence.

This was the first example of programming CRISPR to attack a specified target, and proved that the crRNA determines the specificity of the system. The paper was summarily rejected by *Cell* in April 2012 (without sending it out for external review). A shortened form was submitted to the *Proceedings of the National Academy* and published in September.[15]

Another group, led by Jennifer Doudna at the University of Berkeley, collaborating with Emmanuelle Charpentier at Umeâ in Sweden, found that Cas9 plus crRNA alone do not cleave a target sequence, but the reaction works if another component of the system, a second RNA called tracrRNA, is included. Charpentier had shown the previous year that tracrRNA (*trans*-encoded small RNA) has a 24-base sequence that's complementary to the repeats in crRNA.[16]

Doudna and her collaborators went on to show that the crRNA and tracrRNA could be replaced by combining them to create a joint RNA. They called this a guide RNA, and showed by making five different guide RNAs that it could direct the Cas9 enzyme to cleave DNA with corresponding spacer sequences. Their paper flew into *Science;* submitted at the beginning of June 2012, it was published online by the end of the month.[17]

Perhaps the Doudna–Charpentier paper is more explicit about the potential for gene editing than the Šikšnys paper, but the results are similar. What light does the difference in treatment of the two papers by the journals cast on the conduct of science? For a high-profile journal (such as *Cell, Science,* or *Nature*) the first question is always: would this be interesting if it were true? (For a journal more devoted to a specific field, the question is more likely to be: does this fit into our field of interest?) The second question in either case is: is it true?

The reputation of the authors and their institution(s) should not really have any effect on the first question (and therefore on the decision whether to send the paper for review). The second question should be answered by the data, but it's fair to say there is often more skepticism if the journal editors do not know the authors. It's hard to avoid suspicion that the fact that the Šikšnys paper was first submitted to a journal two months before the Doudna–Charpentier paper, but was published two months after, owes something to the difference in reputation between Vilnius University/DuPont versus the University of Berkeley. Science does not always work as impartially as it should.

At what point did the significance of the system really become clear? "In 2003, I was confident that, at some time, CRISPR will be quoted in the university textbooks and, perhaps, will be taught at high school. Almost two decades later, that's a reality, even though books and teachers talk much more about the technology derived from the native CRISPR systems than on the adaptive immune system of prokaryotes," Francisco Mojica says, a touch ruefully.[18]

With gene editing in sight, from 2012 the pace increased and the sense of competition sharpened, not to say became more aggressive. The challenge was to make the system work in higher organisms as opposed to bacteria.

Jennifer Doudna came into the field in 2006 because she had worked previously on RNA and was asked to collaborate by Jillian Banfield, a microbiologist interested in bacteria growing under extreme conditions, who thought the repeats probably functioned by RNA interference. Two postdocs in the Doudna laboratory, Martin Jinek and Blake Wiedenheft, started working on the system. Their first paper on CRISPR was in 2009, when they characterized the structure of the Cas1 protein. Charpentier was in Vienna, working on tracrRNA at the time, and she had just moved to Umeâ in Sweden when she met Doudna at a conference in 2011 and they decided to collaborate.

Part of the reason why their paper was published so rapidly in *Science* was that Šikšnys and Barrangou had sent an abstract of their paper to Doudna, so she pressured *Science* to publish her paper fast. This may not be one of the more attractive aspects of science, but it is not unusual. At all events, by raising the possibility of genome editing, the paper upped the ante considerably.

The main competitors for Doudna and Charpentier were another set of new players. George Church at Harvard Medical School became interested in the issue, and then his former postdoc, Feng Zhang, worked independently on developing an in vitro system at the Broad Institute at MIT. Zhang was interested in gene editing and viewed CRISPR as potentially a better alternative to any of the available systems. He collaborated with Luciano Marraffini, who had been involved with CRISPR in Erik Sontheimer's laboratory, and was now at Rockefeller University in New York. The first papers from Church and Zhang reporting success with a system in mammalian cells were published in *Science* in February 2013. They were published expeditiously, but not with the breakneck speed of the Doudna–Charpentier paper.[19]

The Doudna laboratory was also working on expressing a CRISPR system in human cells; George Church sent them a copy of his paper when it was accepted at *Science*, and they felt they had to publish their own work immediately, although it had not reached the stage of completion they would have liked. They sent a paper to *eLife*, an online journal with a record of rapid publication, but not (at least yet) in the top rank. Written in only three days, the paper was published (online only) at the very end of January 2013.[20] Demonstrating the inevitability of the development (potentially an important issue in considering patent applications), two further papers from other groups with comparable results were published in January 2013.[21] Jennifer Doudna and Emmanuelle Charpentier were singled out for the Nobel Prize in 2020.

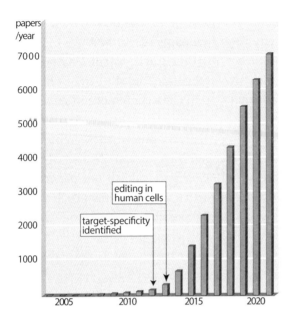

After the potential of the CRISPR system was demonstrated in 2012–2013, the field exploded. Fewer than 10 papers/year were published in 2004–2005, by 2012–2013 the number was in hundreds, and since then it has been in thousands.

The level of competition here was intense, but not unusual for a hot topic. In fact, I would say it is the rule rather than the exception. Authors would call *Cell* all the time to tell us the status of their competitor's papers and urge greater speed in consideration. Sometimes they would call before submission to get an assurance of expeditious handling and reassurance as to when we might publish the paper if the reviews were favorable. (An author who became famous in the DNA sequencing field once called me to suggest a team effort: "we'll do the sequencing and write the papers, and you'll publish them," he proposed.)

Doudna's paper showed that the system makes breaks at sites targeted by the guide RNA. A new buzzword entered the lexicon with the Church and Zhang papers: multiplex, meaning that incorporating several different spacer sequences into the CRISPR array enabled multiple sites to be targeted at once. They went on to show that the breaks made by Cas9 at target sites could be used to stimulate recombination systems that inserted new DNA at these sites. Gene editing had been achieved.

A problem in using a break across both strands of DNA as the initial reaction in editing is that errors can be introduced in putting the DNA back together again. (The double-strand break is a great technique for inactivating a gene to make a "knockout," but greater precision is helpful for editing.) Recent techniques involve linking Cas9 to another enzyme activity that modifies a single base without introducing a break in the DNA.[22]

With the potential for gene editing now obvious, including correcting human diseases, all the principals became involved in companies. Competition was enhanced by a vicious fight over patents.[23] Establishing priority is always an issue in scientific publication: often enough it can be ambiguous, but ambiguity is not permitted in filing for a patent. The point, however, is not the development of antagonism among some of the participants, but the common acceptance of the possibility that the Cas9/guide RNA system could be patented, meaning that researchers who wanted to exploit it might have to pay for its use. Such a possibility would never even have been contemplated at the start of the era of molecular biology.

The history of CRISPR is full of morals about how science works. The first discoveries came from a completely unexpected quarter: curiosity about the purpose of the repeat sequences, followed by investigation of why bacterial cultures got spoiled in yogurt production. There was no way the far-ranging outcome of this apparently mundane situation could have been foreseen.

The young scientists from little-known institutions who were undertaking these studies (more senior figures at major institutions would have regarded the issues as too inconsequential) had difficulties getting their papers accepted by scientific journals. There's potentially a sort of catch-22 in the argument that might have been in the journal editors' minds: if the subject had more importance, it would have been studied by better-known researchers at more important institutions.

When a significant finding made under these circumstances emerges into the light, more established researchers at major institutions do in fact jump on it immediately: its discoverers pretty much lose all hope of being able to stay at the forefront. The big battalions move quickly to take advantage of new opportunities. Big fish eat little fish in science.

How does this story relate to the transition from "hypothesis-driven" science to mining in the database? In a review of CRISPR, Eric Lander (then at MIT) made a vigorous defense of the database. "The history also illustrates the growing role in biology of 'hypothesis-free' discovery based on big data. The discovery ... emerged not from wet-bench experiments but from open-ended bioinformatic exploration of large-scale, often public, genomic datasets."[24]

I'm not sure I agree entirely with this. Francisco Mojica was working on the CRISPR repeats driven more by curiosity than a working hypothesis, and the database (consisting of the published literature) was the means by which he was able to show their general occurrence (and potential importance). Is using the database as a tool to identify the system and its components in principle any different from using a genetic or biochemical method? This led to a hypothesis that the repeats are involved in a bacterial immune system. Investigating that hypothesis led in due course to the discoveries that led to the technology of gene editing. But could this be the last hurrah of old-style science?

CHAPTER
19

EPIGENETICS

The fault, dear Brutus, is not in our stars, But in ourselves... [1]

Cassius, in Julius Caesar

If you aren't confused by quantum mechanics, you haven't really understood it. [2]

Niels Bohr, 1952

Imagine a conversation between Stalin and Lysenko:

Lysenko: Comrade Stalin, I have discovered that chilling the seeds converts Winter Wheat into Spring Wheat.

Stalin: Bravo, Comrade Lysenko, bravo. Practice is the criterion of scientific truth.

Lysenko: We can shorten the time for developing new varieties from ten years to two years.

The theory of inheritance—that the heritable characteristics of an organism or cell are determined solely by its DNA sequence—has ruled biology ever since Mendel. This is not to minimize the importance of the argument about nature versus nurture regarding what effects are hereditary and what effects are environmental, but nature and nurture have always been regarded as *independent* variables.

Potential exceptions to the theory of inheritance go under the general name of *epigenetics*. This is a controversial subject, partly because of misuse of the name. It's a vivid demonstration that what we call something, even in science, can affect how we interpret the data.

At one level, epigenetics describes effects that are exercised at the level of the genome, but that are not directly determined by its sequence.

These occur by fascinating, but intricate, mechanisms that are beyond our scope here. At a deeper level, epigenetics revisits, or perhaps more accurately extends, the nature versus nurture argument by asking whether nurture could in fact influence nature. This challenges the Central Dogma itself. Public consciousness of the issue was shown by an infamous cover of *Time Magazine* in 2010, with the headline "Why your DNA isn't your destiny."

"Since the human genome was sequenced, the term 'epigenetics' is increasingly being associated with the hope that we are more than just the sum of our genes," is how one recent review of epigenetics started.[3] You do not need to go any farther to understand that epigenetics has become a powerful fashion in science, even to the point of sometimes influencing which papers are or are not published.

The influence of epigenetics now extends well beyond genetics. "Following the spectacular rise of epigenetics since the early 2000s, an increasing number of social scientists have called for it to be recognized as an 'interdiscipline,' at the crossroads of the life sciences and the social sciences," is how another review starts. It even introduces the term "social epigenetics" to describe the potential involvement of epigenetics in response to physical or psychological trauma.[4] At its best, this is a stretch; at its worst, a transition from science to pseudoscience, but it emphasizes the importance of understanding the reality and limitations of epigenetics.

The drift of molecular biology in the twentieth century has been reductionist. Through the changes in our views of the capacities of nucleic acid and proteins, everything can be explained in terms of the properties of individual molecules. I do not think many scientists would claim that epigenetics poses a threat to reductionism as such, but it has at times been taken to reopen questions that had been thought to be long-settled, in particular whether environmental effects might be inheritable.

Epigenetics illustrates many of the weaknesses and strengths of science. Although the first indications of epigenetics date from the 1930s, they were impossible to interpret at the time, because there was no context in which to understand them. And then right through the golden age of molecular biology, the overwhelming assumption was that the sequence of DNA is the sole determinant of hereditary properties. But although scientists can be just as prone as anyone else to falling prey to fashion, assumptions in science can do no more than delay the discovery of reality; today epigenetics is an important field. It's a difficult subject, the difficulty being highlighted by how muddled many of its practitioners can become, an illustration of the fact that science can be less than perfect in the pursuit of objectivity.

If biology had laws equivalent to the laws of physics, the first law might be that acquired characteristics cannot be inherited. That law has an importance extending far beyond biology. Attempts to refute that law were the basis for Lamarckism.

Named for Lamarck, who advanced the theory at the start of the nineteenth century, Lamarckism was thoroughly ridiculed through the twentieth century. Putting it in the context of Lamarck's most famous argument, he proposed that when giraffes with short necks stretched to reach higher branches of trees, the necks of the next generation would be longer. That would require stretching to cause a change in the DNA of the sperm or eggs that make the next generation, to code for longer necks. This is implausible. The view of the role of heredity in natural selection today would be that chance mutations in the germ cells (sperm or ova) give some giraffes longer necks, and offspring who inherit the property have an advantage that leads to them taking over the population.

Lamarckism was attractive to Marxist–Leninist doctrine, however, and Lysenko was its main advocate in the Soviet Union through the 1930s until the 1950s. (Trofim Lysenko rejected Mendelian genetics, suppressed other scientists who advocated it, and created the pseudoscience of Lysenkoism, which claimed that the environment could alter heredity.) The devastation that the application of the theory caused in Soviet agriculture—30 million people starved over the period—no doubt added to the sense in the West that it was worse than a scientific error: it was a heresy.

Responding to temperature is a common feature of plant life. Many plants can flower in the Spring only if they are planted and germinated the previous winter. The requirement for a period of prolonged cold before they can flower ensures that they do not attempt to produce seeds until conditions are favorable in the Spring. (For example, they won't respond to a temporary warming during the winter.) This is called *vernalization*.

Arabidopsis thaliana *(mustard cress) is a model organism for plant genetics. It grows to about 10 inches tall.*

Lysenko claimed that this process could be mimicked by chilling seeds. This would mean they could be planted in the Spring for immediate growth, instead of having to be planted the previous Fall. Even more important, he claimed that the property of immediate growth was inherited by the next generation. This would have made it possible to grow crops in more northern areas than had previously been possible.

Largely as the result of work on *Arabidopsis thaliana*, a small plant that became a model for studies of responses to environmental influences, we now understand the basis for vernalization in molecular terms. Genes affecting the process were identified in the 1950s. The most important gene is *FLC (FLOWERING LOCUS C)*. It codes for a protein that turns off the genes that are needed for flowering. While *FLC* is turned on, it's impossible for the plant to flower.

Quis custodiet ipsos custodes? Who guards the guardians? *Arabidopsis* uses a network of many components to sense temperature change, and when the temperature drops, this network turns *FLC* off. The net result is that during cold weather, FLC protein falls to <10% of its starting level. This takes 8–14 weeks, depending on external temperature. This explains why the plant needs a prolonged period of cold before it is able to flower. The plant does not flower at this time, because it is too cold, but the absence of FLC means that it can flower when the weather warms up again.

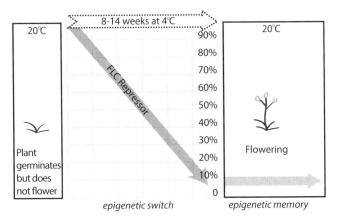

FLC protein prevents flowering. The level of the protein drops over several weeks at cold temperature. FLC remains turned off after the temperature warms up. This allows flowering to occur and is the basis for vernalization.

As *Arabidopsis* grows and cells divide after it has warmed up, the inactive state of the *FLC* gene is perpetuated, so that flowering can occur. We don't need to worry about the details of the mechanism.[5] The important point is that *the gene remembers that it has been turned off.* But when the plant forms seeds, the state of the gene is reset, and it is turned on again, so that flowering is repressed until there has been another period of prolonged cold exposure. Although this refutes any idea that Lysenko could have observed inheritance of characters from one generation to the next, there has all the same been a revisionist school, especially in Russia, to argue that Lysenko might have been right.

As an aside, the mechanism of controlling *FLC* is another demonstration of the conservation of mechanisms throughout evolution, and thus of the fact that information gained from all sorts of organisms, even bacteria, is relevant to humans. The components involved in controlling *FLC* in the plant are related to counterparts that play analogous roles in organisms as distant as fruit flies and humans.[6]

To determine whether epigenetics marks an exception to the dogma of inheritance, we want to ask whether there are situations in which an inherited condition is determined by *some factor other than the inherited sequence of DNA.* What makes control of flowering in *Arabidopsis* an epigenetic effect?

Caroline Dean, from the John Innes Centre in Norwich, who has been a leader in working out the *FLC* circuit, says there are two key reasons why it's epigenetic. "The gene is silenced during the cold but on growth in spring and summer (when it is warm) the gene expression stays off... . Silencing is stored locally, demonstrated by the fact that two copies of *FLC* (in the same cell) can adopt different states of expression."[7]

Here is the heart of the matter. An epigenetic effect created in one set of conditions survives a change of conditions, *and is specific for an individual copy of a gene*. The latter argument is especially important as it draws a firm difference with the conventional control of gene expression, in which both copies of a gene respond in the same way to controlling factors that repress or activate its expression.

Epigenetic effects work by controlling gene expression—determining whether genes are turned on or off—in a way that is self-contained for a particular copy of the gene. That gene maintains its state of expression, which can be different from another copy of the gene, even though both have identical sequences. This contradicts the dogma that the sequence of the gene is solely responsible for its behavior.

The basic principle for controlling the expression of genes was first worked out in bacteria, in the classic work by Jacob and Monod in 1961 (for which they won a Nobel Prize in 1965).[8] The cell makes a regulator (which in principle could be either protein or RNA). Because it is diffusible and can move around the cell, it works on *all* the copies of its target gene. This means that all the copies are turned on or off in unison. Epigenetics defies this principle.

Epigenetics has undergone several changes of meaning since Conrad Waddington first introduced the term in 1942 to describe the way in which the genetic material interacts with its environment. His original definition regards epigenetics as equivalent to what we would now call *development*; this describes the process by which a fertilized egg turns into an organism with many different cell types.

The more specific idea that epigenetics must involve the passage of information through cell division in some form other than the sequence of DNA was formalized in 1994. Then, as epigenetic mechanisms were identified in terms of the modification of DNA or protein components, it

■ CHANGING DEFINITIONS OF EPIGENETICS ■

AUTHOR (YEAR)	DEFINITION
Conrad Waddington (1942)	The branch of biology that studies the causal interactions between genes and their products that bring the phenotype into being.[9]
Robin Holliday (1994)	Nuclear inheritance that is not based on differences in DNA sequence.[10]
Adrian Bird (2007)	The structural adaptation of chromosomal regions so as to register, signal, or perpetuate altered activity states.[11]

came to be identified with the mechanism. These contrasting views can go beyond arguments as to what's in a name into confused logic.

Epigenetics research has exploded, with the number of research papers doubling every three to four years since 2000, reaching around 15,000 papers in 2021. Many (perhaps most) of those papers are not really about epigenetics at all, but may simply be about the control of gene expression.

We don't really need to worry at this point about the mechanism for epigenetic control of *FLC*, except to note that the nomenclature somewhat confuses cause and effect. The process is probably initiated by the addition of a methyl group (a small entity consisting of a carbon atom and three hydrogen atoms) to proteins called histones that are associated with DNA. One possibility is that this could create some sort of templating function that recreates the complex whenever DNA is replicated. (We have a precedent for protein templating effects in the behavior of prions, as discussed in Chapter 15.) The view that histone modifications are a key event is so prevalent that they are often called "epigenetic marks," although this is rather letting the conclusions get ahead of the data.[12]

This is part of the controversy about epigenetics. An alternative view is that histone modifications are simply a consequence of the mechanism for controlling gene expression. It's a running battle in science to distinguish a correlation from cause and effect. Mark Ptashne of the Memorial Sloan Kettering Cancer Center of New York, who has been one of a small group challenging current dogma, says that, "The fashion is that everything is controlled by histone modification. It can be difficult to publish papers with a contrary view."[13]

Mark Ptashne was at Harvard for many years, where early in his career he isolated the repressor of phage lambda (a protein that controls the activity of the virus), a seminal example of a protein that regulates gene action. He has strong views about scientific rigor, and, together with John Greally of Albert Einstein College, led a fierce attack on an excerpt from Mukherjee's book, *The Gene*, which was published in the *New Yorker*, as representing an overly simplified view of epigenetics.[14] This was a sign of the strong opinions—not to say emotions!—that epigenetics engenders.

We have actually known for more than half a century that there are exceptions to the rule that genes with identical sequences behave identically. Cells within a single animal—which by definition must have the same DNA sequence—can form clones with heritable differences. This happens because the two copies of a gene behave differently.

In the major groups of mammals, females have two X chromosomes, and males have one X chromosome and one Y chromosome. This means that females have two copies of each gene carried by the X chromosome, whereas males have only one. If those genes were expressed at a constant level, females would have twice as much of the corresponding proteins as males. This creates a potential problem with dosage.

The problem is solved by inactivating one of the X chromosomes. In 1961, Mary Lyon, a British geneticist, proposed the single X-inactivation hypothesis: one of the two X chromosomes in a female is inactivated at random in each cell early in embryonic development.[15]

The target is the paternal X chromosome in some cells and the maternal X chromosome in other cells. The inactivated state is inherited by all descendent cells.[16] The result is that each cell has only one active copy of each X chromosomal gene, so dosage is the same as in the male. However, if the paternal and maternal copies of the gene are different in a female, the property of a cell may depend on which one is expressed and which one is inactive.

The crucial conclusion from X-inactivation is that *it's impossible to predict the properties of the organism from its DNA sequence*. Because X-inactivation is a *random* process, the behavior of any property controlled by genes on the X chromosome depends on which copy is inactivated in any particular cell in a female. So identical twins (if they are female) could have different characteristics for a property controlled by a gene on the X chromosome.

Beyond the X chromosome, there are cases in which it can make a difference whether a particular gene is inherited from the mother or the father. This is called *imprinting*. The first demonstration that parental chromosomes are not equal came in 1984 when Azim Surani showed that mouse eggs need both a male and female nucleus. "I first used the term genome imprinting in 1984 to describe functional differences between the parental genomes in mammals," Surani says. "The mammalian imprints are heritable but reversible, which entails erasure and re-initiation of the epigenetic imprints in the germline."[17] (Germline cells are sperm or eggs.) This is similar to the situation in *Arabidopsis*: an epigenetic effect is perpetuated in the individual organism, but is reset when the next generation is formed.

The difference from X-inactivation is that the process is not random, but is specific for parental origin. About 15 human diseases are now attributed to imprinting. Whether you get the disease depends on whether a particular gene is inherited from your mother or father. There are probably 100–200 imprinted genes. Imprinted genes sometimes occur in clusters, with several genes controlled together.

The most common cause of imprinting is methylation of DNA to create a heritable state that does not depend on the original sequence of the DNA. (This involves the same type of modification—addition of a methyl group—that controls *FLC* in *Arabidopsis*, but here the target is DNA, not proteins, and the process is better understood.) This is the major exception to the idea that DNA sequence is sufficient as well as necessary to determine behavior.

It was something of a dogma that DNA contains only four nucleotides (A, T, G, C), although the presence of a modified base had been identified well before the discovery of the double helix.[18] Later the modified base turned out to be 5-methylcytosine (where a methyl group is added to cytosine). No one had any idea when the structure of the double helix was worked out that this modification plays a crucial role in inheritance.

Methylation is, of course, a change in the sequence of DNA, so it is not surprising it should affect the behavior of a gene. Its distinctive feature is that the presence of methylcytosine is not determined simply by base-pairing, but a specialized system is required, first to create the methylated cytosine, and then to copy it when DNA replicates. This is called an epigenetic effect because it means that the sequence at this site cannot be predicted from the sequences of the parents.

A role for methylation in controlling gene activity was first suggested in 1985 when Adrian Bird discovered sequences called *CpG islands*, about 30,000 of them, as unmethylated sites in DNA.[19] They are often associated with active genes. Soon after the discovery of the islands, in 1987 two groups found examples of imprinted genes where the expressed copy has unmethylated CpG, whereas the repressed copy is methylated.[20,21]

Now a CpG sequence has a very particular property: it is a *palindrome*, meaning it reads exactly the same on both strands of DNA. Because of the principle that C and G pair only with each other, the CpG on one strand is paired with GpC on the other strand. Remember that the strands run in opposite directions, so in the usual direction of reading, the sequence is the same—CpG—on each strand.

Saying CpG is unmethylated is shorthand for saying it is not methylated on either strand; when it is methylated, the cytosines on both strands have methyl groups. (Methylcytosine base pairs with guanine in exactly the same way as cytosine.) The structure of the palindrome is used to perpetuate the methylated state when DNA replicates, making this a true epigenetic effect. Basically, replication creates daughter helices both of which are methylated only on one strand at the site. An enzyme recognizes this structure and methylates the unmethylated cytosine on each new strand to restore the fully methylated structure of the CpG island. This perpetuates the information.[22]

The imprinted state is reset when sperm or eggs are formed for the next generation.[23] It is easy enough to take the methyl groups off and restore DNA to a pristine sequence: this is done by enzymes called demethylases. The real trick, however, is to restore the pattern that distinguishes between paternal and maternal chromosomes. The key to imprinting is that the chromosomes passed to the next generation have different methylation patterns in eggs or sperm. It remains a mystery how different genes are picked out for methylation in the sperm and the egg.

Imprinting is an enormously clever mechanism for distinguishing between paternal and maternal chromosomes, but it does not transfer environmental effects or pass from one generation to the next.[24] There is much we don't understand about it, but at the end of the day, it is specific to the individual chromosome or gene, making it an epigenetic effect, based on methylation of DNA.[25]

Epigenetics has developed its own language. An epiallele is a copy of a gene that has had a change of state passed on to daughter cells, the epigenome is the set of epialleles, an epigenetic mark is a modification of histone or DNA associated with a change in gene expression, and the histone code describes how modifications of histones may influence gene expression. The problem here is that the terminology carries an implicit assumption that all these events are epigenetic, meaning that they are different from the conventional control of gene expression.

Working out the mechanism of "epigenetic" regulation has led to many novel insights; the quibble is whether applying the terminology of epigenetics more widely is the best way to think about all of these situations, or whether the language is biasing the way we think about the results. Science is supposed to depend on *data*, but here is an illustration that the way we interpret the data can depend on how we describe them.

Epigenetic effects were actually first described in maize by Barbara McClintock in the 1930s, but as this was before the structure of DNA had been elucidated, there was no context within which to understand the results. It was only realized much later that the effects are epigenetic. One of the terms used to describe the effects was paramutation, because they identified a non-Mendelian pattern of inheritance and could not be explained by any simple process of mutation.

Most of McClintock's results were published in the annual report of the Carnegie Institute of Washington, where she worked. When she published a report in 1950 in a conventional journal, the *Proceedings of the National Academy*, it attracted little interest.[26] (McClintock's Nobel Prize did not come until 1983.[27]) It's another demonstration that advances can find widespread appreciation only when the underlying science catches up with them.

Epigenetic effects are better known in plants than in animals, but a classic epigenetic demonstration in mice comes from the behavior of

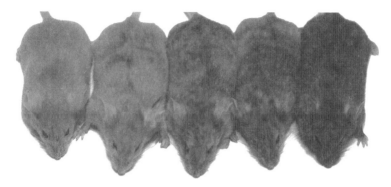

Genetically identical mice can be different. Expression of the A^{vy} gene varies from high enough to give yellow fur (left) to low enough to give the normal striped agouti appearance (right). Courtesy David I.K. Martin.[28]

the *agouti* gene. Animals with the same genetic sequence at the *agouti* gene can appear different. Normal mice, with the *A* gene, have a band of yellow in their fur, between a black base and a black tip. Mutant mice of type *a* have no yellow. Another mutant, A^{vy}, makes completely yellow hair (because the *agouti* gene is overexpressed). A^{vy} is dominant, and *a* is recessive, so mice with one copy of each (i.e., with the genetic constitution A^{vy}/a) should have yellow hair.

In reality, the extent of yellow varies widely. It is inversely correlated with methylation of CpG sequences close to the gene.[29] So depending on how much methylation occurs, mice with the same genetic sequence at the *agouti* gene can appear quite different. However, the quasi-random changes in methylation of DNA that cause these differences in appearance are intrinsic to the genome: they owe nothing to the environment.

The most famous case of epigenetics affecting humans comes from the Dutch Famine (known in the Netherlands as the Hongerwinter). Starvation conditions during the winter of 1944–1945 created a famine in the German-occupied Netherlands. Babies born during this period subsequently showed a range of problems as adults, including obesity.[30] A possible mechanism is revealed by epigenetic changes in the form of reduced methylation at genes that control growth.[31] The theory is that this represents imprinting as a response to fetal deprivation. The story gets contentious with proposals that the effect is passed on to the next generation (i.e., the grandchildren of the mothers who were starved). This is doubtful.

The big question about epigenetics is whether there can be inheritance from one generation to the next. This is now called transgenerational inheritance in the trade. Although there can be questions as to whether transmission of effects within a generation are necessarily epigenetic, there can be no doubt that inheritance across generations that does not depend on the sequence of DNA would truly mark a change in theory. And, of course, unless such effects exist, Lamarckism remains well and truly buried.

More than 40,000 papers have been published on transgenerational epigenetics, but there are remarkably few authentic cases of passage across a generation that is independent of DNA sequence. The most reliable effects are in plants:[32,33] there are also some in the worm *Caenorhabditis elegans*, but few, if any, in higher animals. The most plausible mechanism, as seen in the worm, is the inheritance of a small RNA that affects gene expression.[34]

It is common that when an effect persists across more than one generation in mammals, it dies out over a few generations.[35] The most likely explanation of these effects is that the system for resetting the state of the genes at fertilization does not work completely: some genes escape, allowing an effect to pass to the next generation.[36] The view that environment does not determine heredity prevails.

There's a difference between a field in intellectual ferment when all sorts of ideas are tossed around, some innovative, many wrong, such as in the period at the dawn of molecular biology, from a field in which fashion allows all sorts of idea to be considered without due criticism. The question is whether epigenetics is slipping from the first situation to the second?

Exploding fields are not entirely driven by data: the AIDS epidemic and the COVID pandemic both resulted in scientists rushing into the field with a distinct decline in the signal-to-noise ratio of published papers. What has sometimes cast doubt on epigenetics, if not brought it into disrepute, are unsubstantiated claims for transgenerational inheritance or environmental effects.

Epigenetics is real enough, but it does not apply to every case of gene control. It poses another example of the difficulty science has in assimilating phenomena without any explanatory mechanism. However, "Not knowing the mechanism does not mean we should throw away the concept of epigenetics," Caroline Dean says. The case is best made by the methylation of DNA: here is an effect that distinguishes between different copies of a gene, can be passed through cell division, and can be reset. That is the heart of epigenetics.

Although we are not prisoners of our genes, we are undoubtedly more dependent on hardwired genetic effects than is commonly acknowledged. That dependence is not relieved by epigenetics.

The concept of epigenetics has found its way beyond science into popular parlance. Unfortunately, it is often oversimplified, if not misunderstood. "It is the way plants adapt to the environment," is the way it is often described. Well, yes and no. Without a qualification that the effect is confined to a single generation, this leads directly to Lamarckism, if not back to Lysenko. Whether this is due to misunderstanding of what is admittedly a rather complex concept or to overenthusiastic interpretations of their data by researchers, it supports the view that a little knowledge is a dangerous thing.

The history of epigenetics tells us a good deal about the role of dogmas in science. We started out with a dogma that heredity depends on changes in DNA sequence. Up to this point, the dogma was useful in framing the boundaries for assessing whether epigenetic effects might be an interesting exception or a modification. It is now a fair question whether epigenetics has itself become a dogma, applied to describe effects on the state of gene expression for which it is not necessarily appropriate. At this point, a dogma can become an impediment to progress because it biases the interpretation of new results. Even if science depends on data, and data will always eventually triumph, there can be periods—sometimes quite long periods—when the interpretation is clouded.

STEM CELLS

Every sperm is sacred, Every sperm is great.[1]

Monty Python, 1983

Stem cells belong as the final chapter because they are at the intersection of science with politics and ethics, and illustrate many of the issues of conducting science in a free society. They are an equally hot topic in both science and politics. In biology, there is a great variety of stem cells, but in politics the term is shorthand for human embryonic stem cells.

A stem cell can renew itself and give rise to other types of cells. The early embryo contains stem cells that can generate all the cells of the adult organism; these *embryonic stem cells* are *pluripotent*, in the vernacular of the trade. As tissues develop later, there can be stem cells for individual organs; stem cells for the liver, for example, can form all the cell types of the liver, but not cell types of other organs. Stem cells have enormous medical potential, because in principle they might be able to regenerate replacements for damaged organs. It seems a no-brainer that this should be a priority in biomedical research.

Why can't animals regenerate themselves from cells? To find out, John Gurdon started a series of experiments, first at the University of Oxford, and then at the University of Cambridge, in which he transplanted nuclei from adult frog cells into eggs. This led to a Nobel Prize in 2012.

John Gurdon (now Sir John Gurdon) is an urbane fellow straight out of C.P. Snow's portrayals of life at Oxbridge.[2] He has a liking for college politics and was Master of Magdalene College, Cambridge for almost two decades. He's known for driving around in a flashy Lotus sports car,

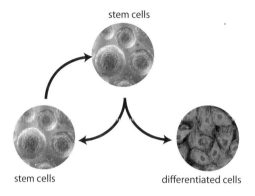

stem cells

stem cells   differentiated cells

Stem cells can perpetuate themselves and can give rise to different cell types. (Some stem cells are pluripotent, others can give rise to a specific set of cell types, such as the epithelial cells shown here.)

but his science is completely sober. His career was scarcely predictable, as he came last in science in his class at Eton, where he was strongly discouraged from becoming a scientist.

His first experiment on frogs was published in 1958 at a time when it was not at all obvious whether all cells in the body have the same set of chromosomes or whether differentiation into adult types is accompanied by loss of some genes. The series of articles developing the system are a classic example of hypothesis-driven science, the question being whether the nucleus or cytoplasm (the outer part of the cell that surrounds the nucleus) defines the potential of the cell to reproduce itself.

The series culminated in a report in 1975 that, when a nucleus is extracted from a skin cell of the frog *Xenopus laevis* (the South African clawed toad), and then inserted into an egg from which the nucleus has been removed, it can form a tadpole. This means that all the genes are present in the skin cell, and none has been irreversibly inactivated.[3]

This could be the basis of a system to clone an individual from a single adult cell, but there were a lot of caveats. Regeneration required more than one step: first, an adult nucleus was implanted into an enucleated egg; then the process was repeated with a nucleus extracted from the early embryo that started to develop. Only 6 out of 129 transfers actually generated tadpoles. None developed into frogs. This is very suggestive that the development of adult cells involves inactivation of genes, that the process requires multiple steps to reverse, and leaves open the question of whether complete reversal is possible.

For the next 20 years, experiments to try to achieve the same feat with cells from mammals were uniformly unsuccessful. It became a dogma that Gurdon's experiments were the exception proving the rule that regeneration of animals would not be possible.

Then came Dolly the sheep, the first cloned mammal.[4] The trick to successful cloning was to induce the donor cells to enter a quiescent phase before the nucleus was extracted. The experiments also worked better when the recipient was an oöcyte (unfertilized egg) and not a fertilized egg. In the case of Dolly, the donor cell providing the nucleus was a breast cell—which is why Dolly was named after Dolly Parton—but other sheep were cloned from other cell types. All this suggested that you have to get exactly the right conditions to allow the donor nucleus to be reprogrammed. (This means resetting the state of gene expression from the donor nucleus to the state characteristic of the egg.)

Over the next decade, it became possible to clone a wide variety of animals by taking a nucleus from an adult cell and effectively creating a new animal from it, although the success rate remained low.[5] (This is called reproductive cloning.) The agricultural implications, that it might be possible to clone animals with desirable traits, were undercut by the fact that most cloned animals had abnormalities, probably resulting from incomplete reprogramming.[6] This added a practical impediment to the ethical considerations for cloning humans.

But there were other more hopeful medical implications. Creating a human embryo with exactly the same genetic makeup as an individual would give access to embryonic stem cells that might be used to regenerate replacement for damaged tissues in that individual. (This

*The South African clawed toad (*Xenopus laevis*) was used for the first experiments on regeneration from nuclei, but they were inconclusive about the potential for cloning.*

Dolly (on the left*) with her surrogate mother. Dolly's white face shows she came from the clone, as she would have a black face if this were her real mother.*

is called therapeutic cloning.) That of course was precisely what regulations introduced in the United States post-Dolly were trying to ban, based on views about the sanctity of human life, as I discuss in Chapter 11.

But the ban applied only to work performed with federal funds. Using other funding sources, Jamie Thomson, at the University of Wisconsin, developed a method in 1998 for culturing cells from early human embryos. Cells came from the blastocyst stage (which has less than 100 cells) of embryos derived from IVF (in vitro fertilization) that had not been used. In effect, the cells were turned into permanent cell lines (called ES for embryonic stem cell).[7] The cell lines would not be useful for cloning or tissue regeneration as such, but they would be enormously useful for studying the underlying processes.

All of this work left open the question of what is responsible for reprogramming a nucleus, so that it loses its restrictions and becomes pluripotent. Epigeneticists see the story as a validation of epigenetics: epigenetic marks have been placed on the chromosome to restrict the activity of the genes. Those marks are usually reset every generation, and somehow the nuclear transfer has mimicked that process, if incompletely.

The truth is more direct. Shinya Yamanaka, at Kyoto University, showed that four *transcription factors* could reverse the state of skin cells and convert them to pluripotent cells. Transcription factors are proteins that turn on genes. Each transcription factor acts on a specific set of genes. Yamanaka got the transcription factors into the cell by incorporating the DNA that codes for them into a retrovirus: by infecting the cells with the retrovirus, the DNA is expressed, and then the transcription factors change the set of genes that are expressed.

First Yamanaka worked with mouse cells; then he showed in 2007 that the same four factors work with human cells.[8] The factors did not come out of the blue: they had been identified as being necessary to maintain pluripotency in the early embryo. In fact, Yamanaka started with 24 candidate factors and narrowed the set down over several years. The new pluripotent cells are called induced pluripotent stem (iPS) cells. Yamanaka shared the Nobel Prize with John Gurdon in 2012.

So there is no moral or ethical issue here: it's possible to generate iPS cells with the same pluripotency as ES cells. Yamanaka went on in 2013 to show that he could use human iPS cells to generate liver buds—small bodies with the structure and function of liver that could be used to assist organ regeneration.[9] It remains to be seen whether this method, with its overt attack on reprogramming, will have better effects than direct nuclear transfer if applied to animals.

And what price epigenetics? At one time, it seemed that epigenetic changes during development, as discussed in Chapter 19, would make it impossible to reverse cell differentiation. Yamanaka's work takes us back to basics. In fact, the mechanism has its origins in the classic work on control of gene expression in bacteria in the 1960s. Since then it's become apparent that regulator proteins may turn genes on or off, and in higher organisms the properties of a cell are determined by its array of these transcription factors.

The basis for Yamanaka's work stands on earlier studies with bacteria, frogs, mice—and human cells. A complete ban on work with human embryonic cells would have undercut the basis for Yamanaka's work, because the ES cells are an essential reference point.

Monty Python's ditty, "Every Sperm Is Sacred," has become the anthem of the beleaguered stem cell community. (The point is sharpened by experiments to convert ES cells into spermatozoa.[10]) The emphasis has shifted—you might even say the community has been saved—by the ability to work on organ-specific stem cells rather than embryonic (pluripotent) cell cells.

The combination of stem cells with gene editing offers the most powerful prospect yet of developing techniques for organ replacement. Creating organ-specific stem cells by the iPS technique eliminates immune rejection by the individual; any genetic basis for a defect in a tissue could be corrected in the stem cells, which could then be used to regenerate the tissue. The brave new world is upon us.

EPILOGUE

EPILOGUE: REDUCTIONISM

The ultimate aim of the modern movement in biology is in fact to explain all biology in terms of physics and chemistry.[1]

Francis Crick, 1966

Anything found to be true of E. coli must also be true of elephants.[2]

Jacques Monod, 1954

Science, at least in the area of molecular biology, has passed from a period driven by hypotheses, indeed sometimes consisting solely of hypotheses because data were so scant, to a transitional period in which hypotheses were more based in data, to the present era in which data trumps all, with the sheer amount of data, even in a PhD thesis, sometimes outrunning the ability of the authors to propose hypotheses.

It was a privilege to be there during the golden age of molecular biology. We are unlikely ever to see its like again, not just because it was a period when fundamental discoveries were fast and furious, but because science was so compact then that you could almost hold it all in your head. Science has so much expanded now that you feel you can only get a handle on small parts of it, and specific technical knowledge is required to understand each area. If this sounds like a lament, I suppose it is, in part.

Particle physics is an older science than molecular biology, but both show a trajectory of successive discoveries from an initial event, accelerating before reaching a point at which the pace slows down. The starting point for physics was the discovery of the atom in the nineteenth century; then the field took off following the discovery of the first particle, the electron, in 1897. The starting point for genetics was the rediscovery of the gene in 1900; then molecular biology took off after the impetus

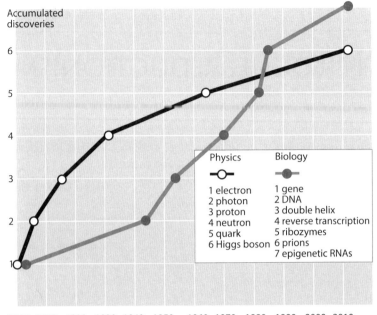

Comparing the discoveries of subatomic particles with discoveries of the fundamental roles of nucleic acids (and prions) shows a similar pattern. The golden age of particle physics was the first half of the twentieth century, when most of the elementary particles were discovered. The golden age of molecular biology was the second half of the twentieth century, when the basic hereditary processes were discovered.

given by the model of the double helix in 1953. Each field had 50 years of discovery from its starting point, followed by a slowing down.

So what is science? Does it have any limitations? A committed reductionist would answer: no, everything can be explained in terms of the laws of physics and chemistry. Certainly molecular biology has provided a full explanation of inheritance and development. Even the apparently mystical aspects of epigenetics have succumbed to analysis.

The universality of DNA as the basis for life prompted no less a person than Francis Crick to advocate directed panspermia—the theory that life on earth originated with seeds sent from another galaxy.[3] Of course, according to Karl Popper's requirement that hypotheses must in principle be falsifiable, this would be pseudoscience rather than science. (Crick admitted as much himself.) Along the lines of fiction, the science fiction serial, *A for Andromeda*, conceived by astronomer Fred Hoyle and broadcast by the BBC in 1961, supposed that an alien intelligence could provide the ability to create life by specifying a sequence of DNA. (The protagonists recognized that a message must correspond to a DNA sequence because it consisted of a series with only four types of information. This shows how deeply the importance of the structure of DNA had already penetrated the public consciousness.)

The serious question is whether the sequence of DNA is sufficient not merely to define a species, but also to recreate it. More specifically, could a cell be created from reading its DNA sequence? (Creating the equivalent of a fertilized egg would effectively create the organism in principle, as the egg has all the information needed for development, if placed in the right environment.)

In the terms of current biology, can all the various "omics" uniquely define a cell? This becomes a quasi-philosophical question at the intersection of epigenetics and DNA sequence.[4] Given that proteins can have a templating function, how far does cell structure depend on preexisting templates of macromolecular assemblies? Could cellular structures, and the cell itself, be created by expressing DNA sequences (allowing that genes must be expressed in a certain order)—or is a preexisting cell (or at least some of its macromolecular assemblies) required?

Every individual human genome has a distinct sequence, but a chemical structure is invariable. Compare the variation in human DNA sequences with the invariant structure of a computer memory chip. Quality control in chip manufacture is all about ensuring absolute consistency in the chip. Quality control in biology is more about feedback loops that compensate for individual variation. There is variance, but not the uncertainty of quantum physics.

Results in biology are not described in terms analogous to the laws of physics. There is no law of gene expression. A gene can be DNA. Or it can be RNA. And a protein can be infectious. Biology is described by mechanisms, shaped for specific tasks. If you want to put it more formally, you could say that the mechanisms of biology must obey the laws of physics, and cannot establish new principles in physics. Molecular biology has been intrinsically reductionist. The very act of proposing generalizations to explain observations implies the use of lower-level rules to explain higher-level interactions.

Physicists may feel that reductionism implies an obligation to require all levels of description to be explained in terms of even lower levels of description, so that ultimately everything would be defined in terms of quantum mechanics. This has led to a heated divide between physicists who are reductionist and those who are antireductionist.[5] I do not believe this conflict is necessarily true of biology. Definitions need simply to be appropriate for the level being investigated. Perhaps the difference is partly because physicists look to define "laws," whereas molecular biologists are content with mechanisms.

We know that DNA replicates by relying on complementary base-pairing between A and T and between C and G. But reductionism does not demand that the reaction occurs in isolation, with DNA spontaneously unwinding to allow adenines to pairs with thymines and cytosines to pair with guanines, let alone that we account for those interactions in terms of quantum mechanics. It is no less reductionist to say that complementary base-pairing is the principle used by other components to accomplish the duplication. This gives perfectly adequate predictability for the reaction.

Perhaps because humans are reluctant to believe they are no more than the sums of their parts, biology has historically offered more scope to oppose reductionism than other sciences. The evolutionary biologist Ernst Mayr, a famous opponent of reductionism, cast doubt in his book of 1987 on whether the discovery of the structure of DNA advanced genetics as such. "The chemical nature of a number of black boxes in the classical genetic theory was filled in, but this did not affect in any way the nature of the theory of transmission genetics."[6]

Although it is true that the structure of DNA does not in itself change our view of Mendelian genetics, this does not take account of the existence of epigenetics. Epigenetics was incomprehensible until mutations (the very foundation of Mendelian genetics) could be interpreted as changes in the sequence of DNA, so that epigenetic effects in turn could be interpreted as modifications imposed on DNA (or chromosomes).

Of course—this being biology—there is a complication. Biology obeys the laws of physics, but it would be hard to apply the laws of physics to predict (for example) the properties of a cell from the properties of its components. Physicist Steve Weinberg, an arch reductionist, admitted that, even in physics, some explanations are "in principle," rather than following from direct application of the relevant laws. He comforted himself by arguing that in principle, we could "calculate anything in chemistry so long as we had a big enough computer and were willing to wait long enough."[7]

No serious scientist would argue any longer that forces beyond the laws of physics and chemistry are required to explain biological systems, but there is an ongoing debate as to validity of the endpoint of reductionism that all higher-level effects can be explained in terms of lower-level effects. The dilemma was very well put by solid-state physicist Philip Anderson of Bell Laboratories, who shared the Nobel Prize in 1977 (physicists have been thinking about these issues for longer than biologists): "The ability to reduce everything to simple fundamental laws does not imply the ability to start from those laws and reconstruct the universe."[8] Anderson was a committed antireductionist.

Contrasted with the reductionist view that nothing can be greater than the sum of its parts, *emergence* argues that more complex properties can emerge from the application of simple rules, but are not necessarily predicted by them. In biology, emergence is most often brought in as a potential explanation for consciousness. In terms of physics, Anderson said that, "at each level of complexity entirely new properties appear."

The atomic scale of the Angstrom (10^{-10} m) defines nuclei and electrons; at the nanometer level (10^{-9} m), the motion of electrons in gold or niobium are the same; moving up three orders of magnitude to the micron level (10^{-6} m), electrons in niobium behave as superconductors at low temperature. (This involves a transition to a state of "broken symmetry," in which, according to Anderson, the whole is "not only more than, but very different, from the sum of its parts.")

I admit that I am disconcerted by the concept of emergence. Like many, if not all, molecular biologists, I have always assumed that the object of science is to explain function in terms of basic principles, and that each level can always be explained by a finer level of resolution. For a reductionist, the concept that new properties can "emerge" at a higher level can be quite alarming. Certainly, the case is far from proven in physics; and experience with biology so far has always supported a reductionist view. Whatever the situation with regard to emergence in physics, it would be premature to accept that it applies in the same way to biology.

The question of whether it might be possible to create a cell from the readout of its DNA sequences poses problems for both reductionism and emergence, because biology depends on time as well as place. From the DNA sequence, we could in principle predict the sequences of all its proteins. But those proteins may fold into the correct structures only in the presence of certain other proteins. The order in which they are expressed may be important. Although that information may be intrinsic to the DNA sequence, its expression depends on correctly defining the starting state (i.e., the condition of a preexisting cell).

Living organisms have (at least) one feature that sets them aside from everything else. They are products of evolution. Their characteristics are not inevitable consequences of the starting state when life was created, but might have been different if the forces of natural selection had been different. If an asteroid had not hit the earth, dinosaurs might still rule.

The situation may be different in physics. According to one view of the Big Bang theory of the universe, everything that followed from the Big Bang could be the inevitable consequence of the operation of the laws of physics (if only we knew what they were). If that were true, it would be the ultimate in reductionism. There are other views, however, supposing that accidental or unpredictable interactions could be decisive.

The starting state for an organism today incorporates not merely the components of the cell but also the history of its evolution. The problem for both reductionism and emergence is that we have now come up against the classic problem of the chicken and the egg. Is it possible to deconstruct the relationship between the starting state of the organism (the egg) and its dependence on the preexisting organism to create that state? And even if it were possible to find conditions that lead to the creation of life-forms, how could we determine whether those conditions had applied billions of years ago, or whether life had been created by some other set of conditions?

There is a view among historians of science that the great insights of Darwinian evolution and Mendelian genetics in biology, and quantum mechanics in physics, are without equal.[9] It in no way minimizes the great accomplishments of the past century to say that Darwin and

Mendel in biology, and Planck in quantum mechanics, started new lineages of thought, and subsequent discoveries have followed in those lineages.

Modern science followed an implicitly reductionist approach until the recent debate about emergence, and it is a fair question whether the problems now remaining in biology and physics will be susceptible to a reductionist approach. It is a sign of the enormous progress of the last century that the term "remaining questions" can be used without irony. If the reductionist approach should now fail, all bets are off, as there is no alternative plan on how to advance. "Emergence" is really no more than an expression of doubt about reductionism: it does not offer a plausible alternative.

Physics seems to be reductionist until you come up against the uncertainty principle of quantum mechanics. If the behavior of an observer can influence whether an electron appears as a wave or a particle, how can that be reduced to an objective observation in the absence of the observer? At this level, physics seems more philosophical than scientific.

Here are two completely different views of the objectives and limits of science. Max Planck, who discovered energy quanta and thereby created quantum physics (winning a Nobel Prize in 1918), said that, "Science cannot solve the ultimate mystery of nature. And that is because, in the last analysis, we ourselves are part of nature and therefore part of the mystery that we are trying to solve."[10] Francis Crick, when he turned to analyzing consciousness in the 1990s, opened his book on the subject by saying, "You're nothing but a pack of neurons."[11] Previously, consciousness had been a bit disreputable as a topic: it took Crick's prestige to make it approachable in scientific (as opposed to philosophical) terms.

The main general issue for the reductionist in biology today is the bootstrap question: can we explain our own consciousness in terms of neural networks? The answer is unknown, and perhaps unknowable, but what is sure is that it will not challenge the laws of physics or chemistry.

Analyses of both the origin of life and the question of consciousness are bedeviled by the impossibility of finding controls for any observations. It is interesting that the only approaches that have been suggested are based on hypothesis-driven science: such questions do not seem to be susceptible to the approach of big data.

Will the question of consciousness mark the end of science? If it became possible to define consciousness at the molecular level, the laments might come true and there could be nothing as interesting left to investigate. If it turns out to be impossible to define consciousness, we may have to acknowledge that science has reached its limits. Either way could be the end of science.

GLOSSARY

α-helix is a common structure formed by a stretch of amino acids in a protein.

Arabidopsis is a small plant (in the cress family) used as a model organism.

Autosomes are all the chromosomes except for the X- and Y-chromosomes involved in determining sex.

Bacteriophage (often known as phage) is a virus that infects bacteria.

Base pair (bp) is the basic unit of DNA.

Caenorhabditis elegans is a nematode worm used as a model organism.

Cas **genes** are the part of the CRISPR system coding for the enzymes that recognize and cut DNA.

Chromosomes contain the genetic material. The number of chromosomes is a characteristic of each organism (see karyotype).

cM (centimorgan) is a unit used to measure genetic linkage between genes.

Codon is a series of three nucleotides that codes for an amino acid (or a stop signal).

CpG island has a sequence of CpG on one DNA strand matching GpC on the other strand. Methylation of CpG islands is associated with turning genes off.

CRISPR is a system that cuts a specific sequence of DNA, determined by a short unique sequence within a series of repeats. Bacteria use it to protect themselves against infection by phages (viruses) that have the target sequence.

Cytoplasm is the part of the cell surrounding the nucleus.

DNA is deoxyribonucleic acid, consisting of two chains, each with a backbone of phosphates linked to sugar (deoxyribose) molecules. Nitrogenous bases are attached to deoxyribose and form the interior, with adenine paired with thymine and cytosine paired with guanine.

DNA sequencing determines the sequence of DNA by breaking strands at specific bases or synthesizing the chain up to a specific base. A series of fragments differing in length by one base, each ending in a specific base, is separated by gel electrophoresis.

Escherichia coli is the standard bacterium used for many experiments in bacterial genetics.

e-Biomed was proposed by Harold Varmus at NIH to establish a free online means of publication for scientific papers.

EcoRI is one of the first restriction enzymes to have been characterized.

Electrophoresis (or gel electrophoresis) is a method used to separate DNA fragments by their length.

ENCODE is a database containing catalogs of DNA sequences, RNA sequences, metabolites, etc.

Epigenetic inheritance refers to a situation in which an effect that does not depend on a change in DNA sequence can be perpetuated through a cell division.

ES cell is an embryonic stem cell, that is, one that can give rise to all the cells of the body.

Eukaryotes have cells that are divided into a nucleus and a cytoplasm. *See also* Prokaryotes.

Exon is a part of a gene that codes for protein. Genes in higher organisms consist of a series of alternating exons and introns.

Fingerprinting (or DNA fingerprinting) uses the fact that every individual has a different pattern of minisatellites to identify individuals or to trace hereditary connections between them.

Gene expression refers to the process of representing a gene in a protein. It involves transcription of DNA into RNA and translation of RNA into protein. It is most often controlled by turning transcription of a gene on or off.

Genetic code is the relationship between the sequence of DNA and the sequence of the protein for which it codes. Three bases in DNA code for each amino acid in protein.

Genome is the complete set of DNA sequences for any organism.

Germline cells are eggs or sperm.

Gigabase (Gb) is a length of 1 billion base pairs.

HHMI is the Howard Hughes Medical Institute, the largest funding source for science in the United States after the NIH.

hnRNA stands for heterogeneous RNA, rather long molecules of RNA found in the nucleus. They are produced by transcribing DNA and are spliced into messenger RNA by removing the sequences corresponding to introns.

HTLV is human T-cell leukemia virus.

HUGO is the Human Genome Organization, which coordinated the international effort at sequencing the human genome.

Hybridoma is a cell producing a single antibody, created by fusing a myeloma with an antibody-producing B cell.

Intron is a part of a gene that does not code for protein and is removed when messenger RNA is formed. Genes in higher organisms consist of a series of alternating exons and introns.

In vitro means working in a test tube (or equivalent) rather than in the living cell or organism.

iPS cell is a pluripotent stem cell created by transferring certain transcription factors into adult cells.

Karyotype is the complete set of chromosomes of an organism.

Kilobase (kb) is a length of 1000 base pairs.

Linkage (or genetic linkage) describes the relationship between genes on the same chromosome. The closer two genes are together, the less frequently they are separated when sperm or eggs are formed, and the smaller the linkage number. It is measured in cM (centimorgans).

Megabase (Mb) is a length of 1 million base pairs.

Metabolomics is the catalog of all the metabolites in a cell.

Methylation is addition of a methyl group (chemical formula CH_3) to a protein or DNA.

Minisatellite is a short repeating sequence found in animal cell DNA that is highly variable. It is the basis for DNA fingerprinting.

mRNA is messenger RNA, produced by transcription of DNA and (in higher organisms) by splicing. Each messenger RNA is translated (by the apparatus of protein synthesis) into a protein.

Myeloma is a cancer of white blood cells. Fusion of a myeloma with a B (antibody-producing) cell can be used to make a hybridoma that produces a specific antibody.

ncRNA means noncoding RNA, an RNA (presumably) involved in controlling gene expression, but not coding for any protein.

Next-generation sequencing is the use of highly automated sequencing machines to sequence DNA.

NIH is the National Institutes of Health. The main campus in Bethesda, Maryland houses many separate institutes, including NCI (National Cancer Institute), and the extramural program is the main source of grants to support science in the United States.

Nucleosome is the basic building block of chromosomes, consisting of 200 bp of DNA wrapped around an octamer of proteins called histones.

Nucleotide is the basic unit of DNA, consisting of a nitrogenous base (a cyclic structure including nitrogen) linked to a 5-carbon sugar, which is linked to a phosphate.

Nucleus is the central part of a (eukaryotic) cell, which includes the chromosomes. It is bounded by a membrane and surrounded by the cytoplasm.

Omics refers to cataloguing the entire set of DNA sequences, RNA sequences, protein sequences, metabolites, etc., in a cell.

PDB is the protein database, a catalog of all proteins whose structures have been solved.

P.I. is a principal investigator, meaning the person who applies for a grant and usually heads a group of post-docs and PhD students.

Phage *See* Bacteriophage.

Plasmid is an independent unit of DNA that can reproduce itself within a bacterium.

Pluripotent stems cells have the capacity to become any cell of the organism.

Postdoc is a postdoctoral researcher, someone who has obtained a PhD but has not yet become an independent investigator.

Preprint is a completed paper that is circulated but has not yet been submitted to a journal for peer review.

Prion is a proteinaceous infectious agent, that is, the infectious form of the PrP protein.

Prokaryotes have cells that consist of a single compartment (there is no nucleus).

PrP is the basic protein subunit of prions. PrP^C is the normal form, and PrP^{SC} is the infectious form.

PubMed is the largest reference database in the world listing scientific papers.

Recombinant DNA is a DNA that has been artificially created by joining DNA from two different sources.

Repressor is a protein that turns off the expression of (usually bacterial) genes.

Restriction enzymes cut DNA at specific short (usually 4- to 6-bp) sequences. They are part of a mechanism by which bacteria defend themselves against infection.

Retrovirus is an RNA virus that uses reverse transcription to produce DNA as part of its life cycle.

Reverse transcription is the process of synthesizing DNA from a template of RNA.

RNA is ribonucleic acid. The chemical difference from DNA is that the sugar is ribose instead of deoxyribose. There are many different types of RNA.

RNA-seq refers to sequencing the RNA complement of a cell.

Sex chromosome is either the X or the Y chromosome; they are involved in determining sex.

Shotgun cloning refers to breaking a genome into random DNA fragments, which after sequences are ordered by using software.

Splicing is the process of removing the sequences of introns from RNA to generate a messenger RNA consisting only of the sequences of exons joined together.

Split gene is a gene that has both exons and introns.

SV40 is simian virus 40, a tumor virus isolated from monkeys.

Transcription is the process of converting a double-stranded sequence of DNA into a single strand of RNA that is identical in sequence with one strand of the DNA.

Transcriptome is the complement of all the RNAs in a cell.

Translation is the process of converting a sequence of mRNA into a sequence of protein. This uses a complex apparatus with the end result that the amino acid sequence of the protein corresponds to the sequence of the mRNA read in triplets.

X chromosome is one of the chromosomes that determine sex. In most mammals, females have two X chromosomes, whereas males have one X chromosome and one Y chromosome.

X-inactivation is the process of inactivating one X chromosome in mammals, so that females have only one active copy of each gene.

NOTES AND REFERENCES

INTRODUCTION: WHAT IS SCIENCE

1. Snow CP. 1993. *The two cultures, and the scientific revolution*, p. 9. Cambridge University Press, Cambridge.

2. Lévi-Strauss C, Eribon D. 1991. *Conversations with Claude Lévi-Strauss*, p. 119. University of Chicago Press, Chicago.

3. Merton RL. 1973. *The normative structure of science* (1942) was reprinted in Merton RK. *The sociology of science*. University of Chicago Press, Chicago.

4. Mostly on the grounds of citing examples of poor behavior in which scientists abused the process; to my mind, this seems more to provide exceptions that prove the rule than to discredit the theory. See Mulkay M. 1969. Some aspects of cultural growth in the natural sciences. *Social Res* **36:** 22–52.

5. The role of mathematics as a science is sometimes questioned because it does not rely on empirical evidence. An important precedent for questioning whether "social sciences" are science comes from Karl Popper's view that work on human behavior is not subject to scientific testing.

6. See note 1, pp. 14–15.

7. Flanders and Swan, *At The Drop of Another Hat*, 1963.

8. See note 1, p. 22.

9. Weinberg S. 1992. *Dreams of a final theory*. Vintage, New York.

10. Kurzweil R. 2005. *The singularity is near*. Penguin, New York.

11. Schrödinger E. 1944. *What is life? The physical aspect of the living cell*. Cambridge University Press, Cambridge.

CHAPTER 1: SCIENCE IN FLUX

1. Medawar PB. 1969. *Induction and intuition in scientific thought*. American Philosophical Society, Philadelphia.

2. The view that science should be driven by hypothesis might be attributed to René Descartes in 1637.

3. Francis Bacon argued in his book *Instauratio Magna*, published in 1620, that science should proceed from inductive reasoning—drawing inferences from observations and data.

4. Popper K. 1934. *The logic of scientific discovery*. Routledge, London.

5. Anderson C. 2008. The end of theory: the data deluge makes the scientific method obsolete. *Wired*, June 23. https://www.wired.com/2008/06/pb-theory

6. Pigliucci M. 2009. The end of theory in science? *EMBO Rep* **10**: 534. doi:10.1038/embor.2009.111; Mazzocchi F. 2015. Could Big Data be the end of theory in science? A few remarks on the epistemology of data-driven science. *EMBO Rep* **16**: 1250–1255. doi:10.15252/embr.201541001

7 Mayer-Schönberger V, Cukier K. 2014. *Big data: a revolution that will transform how we live, work, and think*. Mariner Books, Boston.

8. Davies A, et al. 2021. Advancing mathematics by guiding human intuition with AI. *Nature* **600**: 70–74. doi:10.1038/s41586-021-04086-x

9. Müller U, Ivlev S, Schulz S, Wölper C. 2021. Automated crystal structure determination has its pitfalls: correction to the crystal structures of iodine azide. *Angew Chem Int Ed Engl* **60**: 17452–17454. doi:10.1002/anie.202105666

10. alphafold.ebi.ac.uk

11. Discussion with Demis Hassabis, February 2022.

12. Jumper J, et al. 2021. Highly accurate protein structure prediction with AlphaFold. *Nature* **596**: 583–589. doi:10.1038/s41586-021-03819-2

13. See note 11.

14. de Silva BM, Higdon DM, Brunton SL, Kutz JN. 2020. Discovery of physics from data: universal laws and discrepancies. *Front Artif Intel* **3**: 25. doi:10.3389/frai.2020.00025

15. Demis Hassabis says this is not true of AlphaFold. In fact, he regards fuzzy logic as an outmoded idea that has not contributed much to AI.

16. See note 3.

17. Ross S. 1963. Scientist: the story of a word. *Ann Sci* **18**: 65–85. doi:10.1080/00033796200202722

18. Whewell W. 1840. *The philosophy of the inductive sciences*. Cambridge University Press, Cambridge.

19. Goodstein D. 2007. The method of science. *Nat Phys* **3**: 509. doi:10.1038/nphys689

20. Although scientists are generally dismissive of philosophers of science, the somewhat different views of Karl Popper and Thomas Kuhn are commonplace. Perhaps it is because scientists see Popper as describing the way they would like to do science—testing and falsifying hypotheses—and they recognize the reality of Kuhn's description that much research is adding details, but if/when a paradigm collapses, new ideas replace it.

21. This is often quoted, and it sounds like Feynman, but no one has ever provided a source for it.

22. Sir Peter Medawar drew a distinction between the (nonexistent) "scientific method" and (very much existing) "scientific methodology." See note 1.

23. Philosophers use the term "demarcation" to describe the notion that science is different. Usually they deny it exists.

24. Latour B, Woolgar S. 1986. *Laboratory life. The construction of scientific facts*. Princeton University Press, Princeton, NJ.

25. For a juxtaposition of various views, see Parsons K. 2003. *The science wars: debating scientific knowledge and technology*. Prometheus, New York.

26. For opposing sides of the argument see Gross PR, Levitt N, Lewis MW. 1997. *The flight from science and reason*. Academy of Sciences, New York; Nelkin D. 1996. The science wars: responses to a marriage failed. *Soc Text* **14:** 93–100. doi:10.2307/466846

27. To support their position, constructivists have to go farther into denial of reality. "The process of construction [of facts] involves the use of certain devices whereby all traces of production are made extremely difficult to detect." Latour B, Woolgar S. In Parsons, see note 25.

28. See note 1 in the Introduction, p. 32.

29. Sir Peter Medawar, an immunologist interested in the process of scientific discovery, who ran the National Institute of Medical Research in London, thought the perpetuation of this attitude reflected the British class system. See Medawar PB. 1967. *The art of the soluble*. Methuen, London.

30. Quoted in Crewdson J. 2002. *Science fictions*, p. 240. Little, Brown, New York.

31. Popovic M, Sarngadharan MG, Read E, Gallo RC. 1984. Detection, isolation, and continuous production of cytopathic retroviruses (HTLV-III) from patients with AIDS and pre-AIDS. *Science* **224:** 497–500. doi:10.1126/science.6200935; Gallo RC, et al. 1984. Frequent detection and isolation of cytopathic retroviruses (HTLV-III) from patients with AIDS and at risk for AIDS. *Science* **224:** 500–503. doi:10.1126/science.6200936

32. Rennie D, Yank V, Emanuel L. 1997. When authorship fails: a proposal to make contributors accountable. *J Am Med Assoc* **278:** 579–585. doi:10.1001/jama.278.7.579

33. The roles are Conceptualization, Data curation, Formal analysis, Funding acquisition, Investigation, Methodology, Project administration, Resources, Software, Supervision, Validation, Visualization, Writing—original draft, Writing—review and editing (see credit.niso.org).

34. Wu L, Wang D, Evans JA. 2019. Large teams develop and small teams disrupt science and technology. *Nature* **566:** 378–382. doi:10.1038/s41586-019-0941-9

35. Discussion with David Baltimore, October 2021.

36. Discussion with Tom Cech, November 2021.

CHAPTER 2: DATA MINING

1. Quoted in Gopnik, A. The Porcupine. A pilgrimage to Popper. *New Yorker*, April 1, 2002.

2. Sulston J, Ferry G. 2002. *The common thread*, pp. 33–44. Joseph Henry Press, Washington, DC.

3. Ginsberg J, et al. 2009. Detecting influenza epidemics using search engine query data. *Nature* **457:** 1012–1015. doi:10.1038/nature07634

4. Butler D. 2013. When Google got flu wrong. *Nature* **494:** 155–156. doi:10.1038/494155a

5. Doolittle RF, et al. 1983. Simian sarcoma virus *onc* gene, v-*sis*, is derived from the gene (or genes) encoding a platelet-derived growth factor. *Science* **221:** 275–277. doi:10.1126/science.6304883

6. Longo DL, Drazen JM. 2016. Data sharing. *N Engl J Med* **374**: 276–277. doi:10.1056/NEJMe1516564

7. Based on analysis of research articles in PubMed.

8. Heyneker HL. 2004. "Molecular Geneticist at UCSF and Genentech, Entrepreneur in Biotechnology: an oral history conducted in 2002 by Sally Smith Hughes." Regional Oral History Office, The Bancroft Library, University of California, Berkeley.

9. Roberts L 2001. Controversial from the start. *Science* **291**: 1182–1188. doi:10.1126/science.291.5507.1182a

10. Brown PO, Botstein D. 1999. Exploring the new world of the genome with DNA microarrays. *Nat Genet* **21**: 33–37. doi:10.1038/4462

11. The ENCODE Project Consortium. 2011. A user's guide to the encyclopedia of DNA elements (ENCODE). *PLoS Biol* **9**: 1002046. doi:10.1371/journal.pbio.1001046

12. Hearing of the Subcommittee on Health of the Committee on Energy and Commerce on The Future of Biomedicine: Translating Biomedical Research into Personalized Health Care, December 8, 2021.

13. Data set showing the correlation comes from the COVID-19 Host Genetics Initiative. The link with Neanderthal DNA was shown by Zeberg H, Pääbo S. 2020. The major genetic risk factor for severe COVID-19 is inherited from Neanderthals. *Nature* **587**: 610–612. doi:10.1038/s41586-020-2818-3

CHAPTER 3: PATENTS VERSUS SCIENCE

1. Interview on *See It Now*, CBS, April 12, 1955.

2. Goodfield J. 1981. *An imagined world: a story of scientific discovery*, p. 213. Harper & Row, New York.

3. Sherkow JS, Greely HT. 2015. The history of patenting genetic material. *Ann Rev Genet* **49**: 161–182. doi:10.1146/annurev-genet-112414-054731

4. Diamond v. Chakrabarty. 1980. 447US 303.

5. Cohen SN, Chang AC, Boyer HW, Helling RB. 1973. Construction of biologically functional bacterial plasmids *in vitro*. *Proc Natl Acad Sci* **70**: 3240–3244. doi:10.1073/pnas.70.11.3240

6. Morrow JF, et al. 1974. Replication and transcription of eukaryotic DNA in *Escherichia coli*. *Proc Natl Acad Sci* **71**: 1743–1747. doi:10.1073/pnas.71.5.1743

7. McElheny V. 1974. Gene transplants seen helping farmers and doctors, p. 61. *New York Times*, May 20.

8. Cohen SN. 2013. DNA cloning: a personal view after 40 years. *Proc Natl Acad Sci* **110**: 15521–15529. doi:10.1073/pnas.1313397110

9. Reimers N. 1998. "Stanford's Office of Technology Licensing and the Cohen/Boyer Cloning Patents, an oral history conducted in 1997 by Sally Smith Hughes." Regional Oral History Office, The Bancroft Library, University of California, Berkeley.

10. Boyer HW. 2001. "Recombinant DNA Research at UCSF and Commercial Application at Genentech, an oral history conducted in 1994 by Sally Smith Hughes." Regional Oral History Office, The Bancroft Library, University of California, Berkeley.

11. Hughes SS. 2001. Making dollars out of DNA: the first major patent in biotechnology and the commercialization of molecular biology, 1974–1980. *Isis* **92:** 541–575. doi:10.1086/385281

12. Reimers N. 2021. Don't sabotage the engine of American ingenuity. *The Mercury News*, April 21. https://www.mercurynews.com/2021/04/21/opinion-dont-sabotage-the-engine-of-american-ingenuity

13. Churchill Archives Centre, Milstein Papers, MSTN/C324.

14. See note 16, p. 32.

15. See note 16, p. 25.

16. Quoted in Tansey EM, Catterall PP. 1997. *Technology transfer in Britain: the case of monoclonal antibodies. Wellcome witnesses to twentieth century medicine seminars.* Vol. 1, p. 9. Taylor and Francis, London.

17. Spinks A. 1980. *Biotechnology: report of a joint working party.* Her Majesty's Stationery Office, London.

18. See note 16, pp. 22–23.

19. Wade N. 1980. Inventor of hybridoma technology failed to file for patent. *Science* **208:** 693. doi:10.1126/science.208.4445.693

20. See note 16, p. 28.

21. Roberts L. 1991. Genome patent fight erupts. *Science* **254:** 184–186. doi:10.1126/science.254.5029.184

22. See note 2 in Chapter 2, p. 88.

23. Adler RG. 1992. Genome research: fulfilling the public's expectations for knowledge and commercialization. *Science* **257:** 908–914. doi:10.1126/science.1502557

24. Healy B. 1992. On gene patenting. *N Engl J Med* **327:** 664–668. doi:10.1056/NEJM199208273270930

25. Analysis of data from U.S. Patent Office.

26. See note 22 in Chapter 16, pp. 329–331.

27. See note 2 in Chapter 2, p. viii.

28. This happened in 2005.

29. The E.U. Patent Office issued a Biotechnology Directive in 1998 explicitly permitting patenting of gene sequences.

30. Representing 10 major pharmaceutical companies and 11 biotech companies, but distinctions between them can be blurred.

31. Cook-Degan R, Heaney C. 2010. Patents in genomics and human genetics. *Ann Rev Genomics Hum Genet* **11:** 383–425. doi:10.1146/annurev-genom-082509-141811

32. See note 2 in Chapter 2, p. 90.

33. See note 31.

34. Jensen K, Murray F. 2005. Intellectual property landscape of the human genome. *Science* **310:** 239–240. doi:10.1126/science.1120014

35. King MC. 2014. "The race" to clone *BRCA1. Science* **343:** 1462–1465. doi:10.1126/science.1251900

36. Miki Y, et al. 1994. A strong candidate for the breast and ovarian cancer susceptibility gene *BRCA1*. *Science* **266:** 66–71. doi:10.1126/science.7545954

37. The year the lawsuit was filed, Myriad made $150 million profit on revenues of $360 million, which seems to support the view that its fees were unnecessarily high. See Contreras JL. 2021. *The genome defense.* Workman, New York.

38. More than 100 universities have now agreed to follow a set of principles for licensing patents called the Nine Points, which effectively guarantee that patents will not interfere with basic research.

39. Ogburn WF, Thomas D. 1922. Are inventions inevitable? A note on social evolution. *Polit Sci Q* **37:** 83–98. doi:10.2307/2142320

CHAPTER 4: BIOTECH

1. Boyer HW. 2001. *Recombinant DNA research at UCSF and commercial application at Genentech, an oral history conducted in 1994 by Sally Smith Hughes, Regional Oral History Office, the Bancroft Library*, pp. 95–96. University of California, Berkeley.

2. Parke-Davis in 1902, Lilly in 1911, Upjohn in 1913, Merck in 1933.

3. Lucier P. 2019. Can marketplace science be trusted? *Nature* **574:** 481–485. doi:10.1038/d41586-019-03172-5

4. Quoted in Hughes SS. 2011. *Genentech. The beginnings of biotech*, p. 43. University of Chicago Press, Chicago.

5. See note 4, Hughes, p. 54.

6. Itakura K, et al. 1977. Expression in *Escherichia coli* of a chemically synthesized gene for the hormone somatostatin. *Science* **198:** 1056–1063. doi:10.1126/science.412251

7. Quoted in Wade N. 1980. Cloning gold rush turns basic biology into big business. *Science* **208:** 688–692. doi:10.1126/science.6929110

8. Kenney M. 1986. *Biotechnology: the university-industrial complex*, p. 100. Yale University Press, New Haven, CT.

9. See note 8, pp. 116–121.

10. Rasmussen N. 2014. *Gene jockeys*, p. 164. Johns Hopkins University Press, Baltimore.

11. Separate papers were published reporting the construction of the synthetic gene and its cloning in *E. coli*. Crea R, Kraszewski A, Hirose T, Itakura K. 1978. Chemical synthesis of genes for human insulin. *Proc Natl Acad Sci* **75:** 5765–5769. doi:10.1073/pnas.75.12.5765; Goeddel DV, et al. 1979. Expression in *Escherichia coli* of chemically synthesized genes for human insulin. *Proc Natl Acad Sci* **76:** 106–110. doi:10.1073/pnas.76.1.106

12. See note 10, p. 189.

13. See note 10, p. 130.

14. See note 10.

15. See note 10.

16. Even four decades later, biotechnology has not made insulin cheaply available for all. Its high price became a political campaign issue in 2022: Rojas R. 2022. Price of insulin resonates as issue in Georgia's senate race. *New York Times*, December 1.

CHAPTER 5: THE MYTH OF THE SCIENTIFIC PAPER

1. Schrödinger E. 1944. *What is life? The physical aspect of the living cell.* Cambridge University Press, Cambridge.

2. Medawar P. 1963. Is the scientific paper a fraud? *Listener* **70:** 377–378.

3. Jacob F, Monod J. 1961. Genetic regulatory mechanisms in the synthesis of proteins. *J Mol Biol* **3:** 318–356. doi:10.1016/s0022-2836(61)80072-7

4. Jacob F. 1988. *The statue within. An autobiography* (trans. F. Philip). Unwin Hyman, London.

5. Between 1990 and 1996, 70%–80% of surveyed papers reported data supporting the hypothesis, with the number increasing to 80%–90% from 1997 to 2007. Fanelli D. 2012. Negative results are disappearing from most disciplines and countries. *Scientometrics* **90:** 891–904. doi:10.1007/s11192-011-0494-7

6. Formally this was introduced to describe the formal specification of a post-hoc hypothesis in quantitative analyses.

7. Ritchie S. 2020. *Science fictions.* Metropolitan Books, London.

8. Clinical trials are different, because negative results provide important information about efficacy, and there has been something of a scandal about their suppression by Big Pharma.

9. Letter from Maurice Wilkins to James Watson, October 6, 1966. Available at https://profiles.nlm.nih.gov/spotlight/sc/catalog/nlm:nlmuid-101584582X155-doc

10. Csiszar A. 2016. Troubled from the start. *Nature* **532:** 306–308. doi:10.1038/532306a

11. See Pais A. 1982. *Subtle is the lord: the science and life of Albert Einstein*, p. 494. Oxford University Press, New York.

12. Kennefick D. 2005. Einstein versus the physical review. *Phys Today* **58:** 43–48. doi:10.1063/1.2117822

13. See note 10.

14. Baldwin M. 2018. Scientific autonomy, public accountability, and the rise of peer review in the cold war United States. *Isis* **109:** 538–558. doi:10.1086/700070

15. Baldwin's claim that *Nature* reviewed all submitted papers only from 1973 is wrong. Baldwin M. 2015. *Making nature. The history of a scientific journal.* University of Chicago Press, Chicago.

16. See note 8 in Chapter 12, p. 95.

17. See note 8 in Chapter 12, p. 97.

18. See note 10 in Chapter 4, p. 165.

19. Quoted in Dzeng E. How academia and publishing are destroying scientific innovation: a conversation with Sydney Brenner. *King's Review*, Cambridge, February 2014.

20. See note 10 in Chapter 2.

21. Leading Western publisher bows to Chinese censorship. *New York Times*, November 1, 2017.

CHAPTER 6: REPRODUCIBILITY

1. Pauli W. 1961. *Aufsätze und Vorträge über Physik und Erkenntnistheorie*, p. 94. Vieweg Teubner Verlag, Berlin.

2. Quoted in Hall S. 1987. *Invisible frontiers: the race to synthesize a human gene*, p. 242. Atlantic Monthly Press, New York.

3. Baker M. 2016. Is there a reproducibility crisis? *Nature* **533**: 452–453.

4. More than 50% of respondents believed that >70% of results are reproducible.

5. Errington T, et al. 2021. Reproducibility in cancer biology: challenges for assessing replicability in preclinical cancer biology. *eLife* **10**: e67995. doi:10.7554/eLife.67995

6. Errington T, et al. 2021. Investigating the replicability of preclinical cancer biology. *eLife* **10**: e71601. doi:10.7554/eLife.71601

7. Prinz T, Schlange T, Asadullah K. 2011. Believe it or not: how much can we rely on published data on potential drug targets? *Nat Rev Drug Disc* **10**: 712–713. doi:10.1038/nrd3439-c1; Begley CG, Ellis LM. 2012. Raise standards for preclinical cancer research. *Nature* **483**: 531–533. doi:10.1038/483531a

8. Hay M, Thomas DW, Craighead JL, Economides C, Rosenthal J. 2014. Clinical development success rates for investigational drugs. *Nat Biotechnol* **32**: 40–51. doi:10.1038/nbt.2786

9. The most egregious manipulations to present clinical trial results in a more favorable light by Pfizer, Merck, and Schering-Plough were reported by Saul S. 2008. Experts conclude Pfizer manipulated studies. *New York Times*. October 8. For a wider study, see McGauran N, et al. 2010. Reporting bias in medical research—a narrative review. *Trials* **11**: 37. doi:10.1186/1745-6215-11-37

10. Camerer CF, et al. 2016. Evaluating replicability of laboratory experiments in economics. *Science* **351**: 1433–1436. doi:10.1126/science.aaf0918

11. Camerer CF, et al. 2018. Evaluating the replicability of social science experiments in *Nature* and *Science* between 2010 and 2015. *Nat Hum Behav* **2**: 637–644. doi:10.1038/s41562-018-0399-z

12. Open Science Collaboration. 2015. Estimating the reproducibility of psychological science. *Science* **349**: aac4716. doi:10.1126/science.aac4716

13. For contrary views, see Ioannidis JP. 2005. Why most published research findings are false. *PLoS Med* **2**: e124. doi:10.1371/journal.pmed.0020124; and Fanelli D. 2018. Is science really facing a reproducibility crisis, and do we need it to? *Proc Natl Acad Sci* **115**: 2628–2631. doi:10.1073/pnas.1708272114

14. Taubes G. 1993. *Bad science: the short life and weird times of cold fusion*. Random House, New York.

15. Courtesy Richard Myers. https://dukespace.lib.duke.edu/dspace/handle/10161/7721

16. Sulston J, Ferry G. 2002. *The common thread*, p. 229. Joseph Henry Press, Washington, DC.

17. Venter was quoted in the news media as saying, "We said that once we had finished sequencing the genome we would make it available to the scientific community for free, and we will be doing that on Monday morning at 10 a.m.," but this was misleading given the conditions for access. http://www.cnn.com/2001/HEALTH/02/09/genome.results/index.html

18. King J. 2005. Where the future went. *EMBO Rep* **6:** 1012–1014. doi:10.1038/sj.embor.7400553

19. Read RJ, et al. 2011. A new generation of crystallographic validation tools for the protein data bank. *Structure* **19:** 1395–1412. doi:10.1016/j.str.2011.08.006

20. Minor W, et al. 2016. Safeguarding structural data repositories against bad apples. *Structure* **24:** 218–220. doi:10.1016/j.str.2015.12.010

CHAPTER 7: PUBLISH OR PERISH

1. Quoted in *The Guardian*, December 6, 2013.

2. Of course, the answer can depend on how you define "sound." If it is defined as the transmission to the ear of vibrations in the air generated by the falling tree, then sound does not exist in the absence of the observer. This is a quibble that would, however, be recognized by quantum physicists who are wont to question whether effects exist in the absence of observers. I would argue that at this level, science has degenerated into philosophy.

3. The National Library of Medicine states that about 30,000 records are included in the PubMed journal list.

4. Number of scientific papers each year included in PubMed.

5. The number of published papers doubled between 2006 and 2020, whereas funding for the NIH (for example) increased 40% in the same period. NIH Office of Budget. Appropriations History by Institute/Center (1938 to Present).

6. Communicated to the President. 1744. An account of some experiments, lately made in Holland, upon the fragility of unannealed glass vessels. *Philos Trans R Soc* **488:** 505–516.

7. Haines J. 1988. *Maxwell*, p. 137. Houghton Mifflin, Boston.

8. Larivière V, Haustein S, Mongeon P. 2015. The oligopoly of academic publishers in the digital era. *PLoS ONE* **10:** e0127502. doi:10.1371/journal.pone.0127502

9. Hamilton DP. 1990. Publishing by—and for?—the numbers. *Science* **250:** 1331–1332. doi:10.1126/science.2255902

10. Pendlebury DA. 1991. Science, citation, and funding. *Science* **251:** 1410–1411. doi:10.1126/science.251.5000.1410-b

11. Based on analysis of 8 million papers published between 2000 and 2015 available in PubMed in December 2021. Medicine actually came off worse than science, with ~25% uncited compared with 10% for biology. The PubMed USA data set has 3.75 million papers, Medline has 7.5 million papers, and the NIH data set has 630,000 papers.

SOURCE	NEVER CITED (%)	CITED ONLY BY AUTHOR (%)	CITED >4 TIMES (%)
PubMed	20.9	2.4	48.6
PubMed USA	18.7	2.2	54.3
Medline	19.4	2.4	50.3
Funded by the NIH	2.5	2.0	84.1

12. This may underestimate its importance, as eventually a highly cited paper may become so well-known that investigators no longer feel a need to cite it; this has been called citation obliteration.

13. Garfield E. 1955. Citation indexes for science: a new dimension in documentation through association of ideas. *Science* **122**: 108–111. doi:10.1126/science.122.3159.108

14. The usual period for calculating the impact factor is two years, but this is arbitrary, and other periods could be used.

15. Verma IM. 2014. Simplifying the direct submission process. *Proc Natl Acad Sci* **111**: 14311. doi:10.1073/pnas.1417688111

16. Davis PM. Comparing the citation performance of *PNAS* papers by submission track. www.biorxiv.org/content/10.1101/036616v2.

17. Since its establishment in 2012, *eLife* has published more than 11,000 papers, which have cited more than 250,000 other papers. Analysis is based on research articles published from 2012 to 2021.

18. Ortega y Gasset J. 1932. *The revolt of the masses*. Norton, New York.

19. Cole JR, Cole S. 1972. The Ortega hypothesis. *Science* **178**: 368–375. doi:10.1126/science.178.4059.368; Bornmann L, de Moya Anegón F, Leydesdorff L. 2010. Do scientific advancements lean on the shoulders of giants? A bibliometric investigation of the Ortega hypothesis. *PLoS ONE* **5**: e13327. doi:10.1371/journal.pone.0013327

20. Elsevier 4492; Springer, 3763; Taylor & Francis, 2913; John Wiley, 2737; Sage, 1000. From the publishers' websites.

21. Analysis of research articles published in 2018 from PubMed.

22. Based on citations reported in PubMed by December 2021 for papers published in 2016. The Cell Press group, Lancet group, and Nature group were excluded from the major publishers.

PUBLISHER	NEVER CITED OR CITED ONLY BY AUTHOR (%)	CITED >4 TIMES (%)
Sage	36.6	27.3
Kluwer	36.3	30.7
Cambridge University Press	36.3	29.6
Taylor & Francis	32.7	31.8
Springer	32.6	34.2
Wiley	30.4	39.8
Elsevier	30.2	37.3
Nature Pub Group	18.2	60.0
PLoS	9.1	79.6
Cell Press	7.3	82.0

23. Springer-Verlag offers nothing for free. The cost of buying an individual article, even one that is only of historical interest, is prohibitive, and the e-mail symbol next to the author's name does not in fact have the e-mail address, so to contact the author by e-mail you have to find another source.

24. https://retractionwatch.com/2021/11/04/springer-nature-geosciences-journal-retracts-44-articles-filled-with-gibberish/#more-123459

25. https://link.springer.com/article/10.1007/s12517-021-08804-7

26. Angell M, Kassirer JP. 1991. The Ingelfinger rule revisited. *N Engl J Med* **325:** 1371–1373. doi:10.1056/NEJM199111073251910

27. Discussion with Marcia Angell, December 2021.

28. Garfield E. 1982. *ISI atlas of science: biochemistry and molecular biology 1978/80*. Institute for Scientific Information, Philadelphia.

29. The volume presented a list of 11 contributors, but the individual minireviews were anonymous, although each was reviewed by an unnamed "international expert in the particular field."

30. Lewin B. 1982. A little knowledge is a fearsome thing. *Cell* **30:** 661–665.

31. Zuckerman H. 1977. *Scientific elite*. The Free Press, New York.

32. "The prevailing pragmatism forced upon the academic group is that one must write something and get it into print. Situational imperatives dictate a 'publish or perish' credo within the ranks"; Wilson L. 1942. *The academic man: a study in the sociology of a profession*, p. 197. Oxford University Press, New York.

33. See note 1.

Chapter 8: E-Science

1. Quoted in Leary WE. 2003. Measure calls for wider access to federally financed research. *New York Times*, June 26.

2. Quoted in Poynder R. 2005. Interview with Vitek Tracz: essential for science. *Information Today* **22**. https://www.infotoday.com/it/jan05/poynder.shtml

3. "Almost" because there is the issue of permanence: When a library has a physical copy of the material, it can ensure that it will always be available, but websites can disappear. Indeed, some online journals started as the result of early enthusiasm have done just that, taking their data with them. That could be dealt with, of course, by depositing a copy of the site elsewhere, much as books are deposited with Library of Congress for copyright.

4. *PNAS* is a pioneer leading the way to an electronic-only world. It became online-only in 2019, although print format is still available by request.

5. Varmus H. 1999. *E-BioMed: a proposal for electronic publication in the biomedical sciences*. NIH preprint 04.99 doc. National Institutes of Health, Bethesda, MD.

6. Cobb M. 2017. The prehistory of biology preprints: a forgotten experiment from the 1960s. *PLoS Biol* **15:** e2003995. doi:10.1371/journal.pbio.2003995

7. Varmus H. 2009. *The art and politics of science*, p. 270. Norton, New York.

8. Smith R. 2006. Lapses at the *New England Journal of Medicine. JR Soc Med* **99:** 380–382. doi:10.1177/014107680609900802

9. Relman AS. 1999. The NIH "e-BioMed" proposal—a potential threat to the evaluation and orderly dissemination of new clinical studies. *N Engl J Med* **340:** 1828–1829. doi:10.1056/NEJM199906103402309

10. plos.org/open-letter/

11. See note 7, p. 250.

12. Discussion with Harold Varmus, January 2022.

13. liblicense.crl.edu/ListArchives/0310/msg00048.html

14. https://www.whitehouse.gov/wp-content/uploads/2022/08/08-2022-OSTP-Public-Access-Memo.pdf

15. https://www.embo.org/features/the-publishing-costs-at-embo/

16. Schekman R. 2013. How journals like *Nature, Cell* and *Science* are damaging science. *The Guardian*, December 9.

17. Callaway E. 2016. Biology's big funders boost *eLife. Nature* **540:** 14–15. doi:10.1038/534014a

18. Schekman R. 2019. Scientific publishing: progress and promise. *eLife* **8:** e44799. doi:10.7554/eLife.44799

19. Eisen MB, et al. 2020. Peer review: implementing a "publish, then review" model of publishing. *eLife* **9:** e64910. doi:10.7554/eLife.64910

20. Grudniewicz A, et al. 2019. Predatory journals: no definition, no defence. *Nature* **576:** 210–212. doi:10.1038/d41586-019-03759-y

21. https://beallslist.net

22. The term was probably introduced by Jeffrey Beall, a librarian at the University of Colorado, Denver. See Beall J. 2012. Predatory publishers are corrupting open access. *Nature* **489:** 179. doi:10.1038/489179a

CHAPTER 9: FUNDING

1. Planck M. 2014. *Scientific autobiography: and other papers.* Philosophical Library.

2. National Science Foundation, Growth in Federal Research Obligations for Life Sciences between 1970 and 2020.

3. von Hippel T, von Hippel C. 2015. To apply or not to apply: a survey analysis of grant writing costs and benefits. *PLoS ONE* **10:** e0118494. doi:10.1371/journal.pone.0118494

4. Velazquez JLP. 2019. *The rise of the scientist-bureaucrat.* Springer, Switzerland.

5. report.nih.gov/nihdatabook

6. NIH data book.

7. Capecchi MR. 2008. A Nobel lesson: the grant behind the prize. Response. *Science* **319:** 900–901.

8. See note 4.

9. Kaiser J. NIH plan to reduce overhead payments draws fire. *Science* **356:** 893. doi:10.1126/science.356.6342.893

10. Parr C. 2014. *Times Higher Education,* December 3.

11. Azoulay P, Graff Zivin JS, Manso G. 2011. Incentives and creativity: evidence from the academic life sciences. *RAND J Econ* **42:** 527–554. doi:10.1111/j.1756-2171.2011.00140.x

12. Pier EL, et al. 2018. Low agreement among reviewers evaluating the same NIH grant applications. *Proc Natl Acad Sci* **115:** 2952–2957. doi:10.1073/pnas.1714379115

13. Fang FC, Bowen A, Casadevall A. 2016. NIH peer review percentile scores are poorly predictive of grant productivity. *eLife* **5:** e13323. doi:10.7554/eLife.13323

14. Alberts B, Kirschner MW, Tilghman S, Varmus H. 2014. Rescuing US biomedical research from its systemic flaws. *Proc Natl Acad Sci* **111:** 5773–5777. doi:10.1073/pnas.1404402111

15. Quoted in Varmus H, Marshall E. 1993. Varmus: the view from Bethesda. *Science* **262:** 1364–1365. doi:10.1126/science.8248775

16. Fang FC, Casadevall A. 2016. Research funding: the case for a modified lottery. *mBio* **7:** e00422-16. doi:10.1128/mBio.00422-16

17. Yeats WB. 1919. *The second coming.*

18. Impact factors are stated as a specific factor in considering tenure at 40% of research-based universities in the United States and Canada. McKiernan EC, et al. 2019. Meta-Research: use of the Journal Impact Factor in academic review, promotion, and tenure evaluations. *eLife* **8:** e47338. doi:10.7554/eLife.47338

CHAPTER 10: FRAUD

1. Delbrück M. Interview by Carolyn Harding. Pasadena, California, July 14–September 11, 1978. Oral History Project, Caltech Archives. http://resolver.caltech.edu/CaltechOH:OH_Delbruck_M.

2. See note 15 in Chapter 9.

3. Holton G. 1978. *The scientific imagination: case studies,* pp. 25–83. Cambridge University Press, Cambridge.

4. The most probable reason for ignoring them was that they were preparatory: Segerstråle U. 1995. Good to the last drop? Millikan stories as "canned" pedagogy. *Sci Eng Ethics* **1:** 197–214. doi:10.1007/BF02628797; Goodstein D. 2010. *On fact and fraud.* Princeton University Press, Princeton, NJ.

5. Westfall RS. 1981. *Never at rest: a biography of Isaac Newton.* Cambridge University Press, Cambridge.

6. Levi-Montalcini R. 1988. *In praise of imperfection.* Basic Books, New York.

7. *Responsible science: ensuring the integrity of the research process.* 1992. National Academy Press, Washington, DC.

8. Babbage C. 1830. *Reflections on the decline of science in England, and on some of its causes.* London. (Most of the book consisted of complaints about the Royal Society, of which he was a member.)

9. Koshland DE. 1987. Fraud in science. *Science* **235**: 141. doi:10.1126/science.3798097

10. Based on retractions in the PubMed database as of December 2021.

11. Some journals publish so few that the absence seems more likely to represent editorial reluctance than higher accuracy in the editorial process.

12. China = 44.1%, rest of Asia = 12.7%, United States = 8.4%, Europe = 6.5%, according to www.retractionwatch.com, January 2022. This applies only to papers in which all authors are from a single country.

13. Calculated as the percentage of papers retracted for journals in biology, medicine, or physics.

14. Plagiarism is different from other categories because it does not change the validity of data but misattributes it.

15. Budd JM, Sievert M, Schultz TR, Scoville C. 1999. Effects of article retraction on citation and practice in medicine. *Bull Med Libr Assoc* **87**: 437–443; Fang FC, Steen RG, Casadevall A. 2012. Misconduct accounts for the majority of retracted scientific publications. *Proc Natl Acad Sci* **109**: 17028–17033. doi:10.1073/pnas.1212247109

16. Analysis of the Retraction Watch database, January 2022.

17. Data from www.retractionwatch.com, January 2022.

FRAUD		
Duplication	5218	24%
Plagiarism	3806	18%
Faking data	2150	10%
Fake peer review	1140	5%
ERRORS		
Error in data	3855	18%
Concerns about data	1743	8%
Withdrawn	1725	8%
Unknown	1345	6%
Not reproducible	698	3%
Ethics	673	3%
Total	21655	

18. See Chapter 5 for the calculation of frequency of retractions.

19. See note 12.

20. The Retraction Watch leaderboard (retractionwatch.com/the-retraction-watch-leaderboard/) lists the authors with the most retractions. This shows that 31 authors have 1355 retractions, so they account for almost 5% of all retractions.

21. Bornemann-Cimenti H, Szilagyi IS, Sandner-Kiesling A. 2016. Perpetuation of retracted publications using the example of the Scott S. Reuben case: incidences, reasons and possible improvements. *Sci Eng Ethics* **22**: 1063–1072. doi:10.1007/s11948-015-9680-y

22. Based on 22,341 retractions of research articles in the Retraction Watch database as of January 2022. The frequency of retraction stated in the legend is calculated relative to the number of papers in PubMed since 2010.

23. There's a popular view that papers in *Nature*, *Science*, and *Cell* are more likely to be fraudulent and retracted because they are more cutting-edge, but this is not true, at least not of the (admittedly limited) data set in note 22.

24. Fang FC, Casadevall A. 2011. Retracted science and the retraction index. *Infect Immun* **79**: 3855–3859. doi:10.1128/IAI.05661-11

25. My own analysis of 12,531 retractions published in 2538 journals indexed in PubMed from 2010 to 2021 suggests that the correlation between impact factor and retraction frequency is −0.14.

26. The median is five papers and/or grants in the current ORI list.

27. Bik EM. 2016. The prevalence of inappropriate image duplication in biomedical research publications. *mBio* **7**: e00809-16. doi:10.1128/mBio.00809-16

28. Van Noorden R. 2022. Journals adopt AI to spot duplicated images in manuscripts. *Nature* **601**: 14–15. doi:10.1038/d41586-021-03807-6

29. COPE (Committee on Publication Ethics). www.publicationethics.org

30. Stroebe W, Postmes T, Spears R. 2012. Scientific misconduct and the myth of self-correction in science. *Perspect Psychol Sci* **7**: 670–688. doi:10.1177/1745691612460687

31. https://ori.hhs.gov/historical-background

32. https://ori.hhs.gov/content/case-summary-downs-charles-a

33. Granted that the reports include cases of scientific misconduct that can vary from outright fraud to lesser offenses, it's still surprising to see a four-year probationary period applied as the penalty for cases that use the words "fraud" and "fabricating."

Chapter 11: Politics and Ethics

1. UCLA conference.

2. Lander E, et al. 2019. Adopt a moratorium on heritable genome editing. *Nature* **567**: 165–168. doi:10.1038/d41586-019-00726-5

3. Mertz JE, Davis RW. 1972. Cleavage of DNA by R_1 restriction endonuclease generates cohesive ends. *Proc Natl Acad Sci* **69**: 3370–3374. doi:10.1073/pnas.69.11.3370

4. Jackson DA, Symons RH, Berg P. 1972. Biochemical method for inserting new genetic information into DNA of simian virus 40: circular SV40 DNA molecules containing lambda phage genes and the galactose operon of *Escherichia coli*. *Proc Natl Acad Sci* **69**: 2904–2909. doi:10.1073/pnas.69.10.2904

5. Chang ACY, Chen SN. 1974. Genome construction between bacterial species in vitro: replication and expression of *Staphylococcus* plasmid genes in *Escherichia coli*. *Proc Natl Acad Sci* **71**: 1030–1034. doi:10.1073/pnas.71.4.1030

6. https://dnalc.cshl.edu/view/15017-reaction-to-outrage-over-recombinant-DNA-Paul-Berg.html

7. Mukherjee S. 2016. *The gene. An intimate history*, p. 226. Scribner, New York.

8. Hanna KE. 1991. *Biomedical politics. Institute of Medicine (US) Committee to Study Decision Making*, p. 268. National Academies Press (US), Washington, DC.

9. The Gordon Conferences were at the time among the most prestigious scientific meetings in the United States. Restricted to a relatively small number of invited participants, they were held at shabby boarding schools in New Hampshire.

10. Singer M, Soll D. 1973. Guidelines for DNA hybrid molecules. *Science* **181**: 1114. doi:10,1126/science.181.4105.1114

11. David Baltimore, James Watson, Dan Nathans, Sherman Weissman, Norton Zinder, and Richard Roblin.

12. Berg P, et al. 1974. Potential biohazards of recombinant DNA molecules. *Science* **185**: 303. doi:10.1126/science.185.4148.303

13. Berg P, Baltimore D, Brenner S, Roblin RO III, Singer MF. 1975. Asilomar conference on recombinant DNA molecules. *Science* **188**: 991–994. doi:10.1126/science.1056638

14. *Harvard Crimson*, May 17, 1977.

15. Letter to Senator Harrison Schmitt of the Subcommittee on Science, Technology, and Space, January 5, 1979.

16. Greve JM, et al. 1989. The major human rhinovirus receptor is ICAM-1. *Cell* **56**: 839–847. doi:10.1016/0092-8674(89)90688-0; Staunton DE, et al. 1989. A cell adhesion molecule, ICAM-1, is the major surface receptor for rhinoviruses. *Cell* **56**: 849–853. doi:10.1016/0092-8674(89)90689-2

17. Wilmut I, et al. 1997. Viable offspring derived from fetal and adult mammalian cells. *Nature* **385**: 810–813. doi:10.1089/clo.2006.0002

18. Stock G, Campbell J. 2000. *Engineering the human germline: an exploration of the science and ethics of altering the genes we pass to our children.* Oxford University Press, New York.

19. Baltimore D, et al. 2015. A prudent path forward for genomic engineering and germline gene modification. *Science* **348**: 36–38. doi:10.1126/science.aab1028

20. There were some technical questions as to whether *CCR5* had in fact been completely eliminated, but there was no doubt that gene editing had been applied to the human germline.

21. See note 24 in Chapter 18.

22. Reviews are not necessarily peer-reviewed as they do not include original data.

23. See note 2.

24. National Academy of Sciences. 2020. *Heritable human genome editing.* The National Academies Press, Washington, DC.

25. Varmus recounts the history in note 7 in Chapter 8, pp. 150–151.

26. See note 7 in Chapter 8, p. 206.

27. Bush V. 1945. Science, the endless frontier. *Trans Kansas Acad Sci* **48**: 231–264. doi:10.2307/3625196

28. Stokes DE. 1997. *Pasteur's quadrant: basic science and technological innovation*, p. 98. Brookings Institution Press, Washington, DC.

CHAPTER 12: MENDEL'S GARDEN

1. Mendel G. 1865. Experiments on plant hybrids. Translation: Abbott S, Fairbanks DJ. 2016. *Genetics* **204**: 407–422. doi:10.1534/genetics.116.195198

2. Stent GS. 1972. Prematurity and uniqueness in scientific discovery. *Sci Am* **227**: 84–93. doi:10.1038/scientificamerican1272-84

3. Smooth:wrinkled: 253 plants produced 7324 seeds with 5474 smooth and 1850 wrinkled, giving a ratio of 2.96:1. Yellow:green: 258 plants produced 8023 seeds, with 6022 yellow and 2001 green, giving a ratio of 3.01:1.

4. Franklin A, Edwards AW, Fairbanks DJ, Hartl DL. 2008. *Ending the Mendel–Fisher controversy*. University of Pittsburgh Press, Pittsburgh.

5. "Gene" was first used to describe the unit of Mendelian genetics by a Danish botanist, Wilhelm Johannsen, in 1909. It came into general use when *Drosophila* geneticist Thomas Hunt Morgan published his book in 1915, *The Mechanism of Mendelian Heredity*. The term "genetics" had been introduced previously by William Bateson, in 1905.

6. Schrödinger E. 1944. *What is life? The physical aspect of the living cell.* Cambridge University Press, Cambridge.

7. From a letter Avery wrote to his brother in 1943, quoted in Dubos RJ. 1976. *The professor, the Institute, and DNA*, pp. 218–219. The Rockefeller University Press, New York.

8. Watson JD. 2000. *A passion for DNA. Genes, genomes, and society*, p. 9. Cold Spring Harbor Laboratory Press, Cold Spring Harbor, NY.

9. Hargittai I. 2009. The tetranucleotide hypothesis: a centennial. *Struct Chem* **20**: 753–756. doi:10.1007/s11224-009-9497-x

10. The demonstration that RNA can also provide genetic material did not come until 1956: Fraenkel-Conrat H. 1956. The role of the nucleic acid in the reconstitution of active tobacco mosaic virus. *J Am Chem Soc* **78**: 882–883. doi:10.1021/ja01585a055

11. The significance of RNA was only appreciated because Francis Crick proselytized for the importance of the result: See note 16 in Chapter 13, p. 239.

CHAPTER 13: THE DOUBLE HELIX

1. Levene PA. 1917. The chemical individuality of tissue elements and its biological significance. *J Am Chem Soc* **39**: 828–839. doi:10.1021/ja02249.a037

2. Watson JD, Crick FHC. 1953. Molecular structure of deoxypentose nucleic acids. *Nature* **171**: 737–738. doi:10.1038/171737a0

3. Watson J. 1968. *The double helix*, p. 15. Weidenfeld and Nicolson, London.

4. Bragg WL. 1965. First stages in the X-ray analysis of proteins. *Rep Prog Phys* **28**: 1–16.

5. See note 3, p. 172.

6. Paterlini M. 2003. History and science united to vindicate Perutz. *Nature* **424**: 127. doi:10.1038/424127a

7. See note 3, p. 165.

8. Astbury WT, Bell FO. 1938. Some recent developments in the X-ray study of proteins and related structures. *Cold Spring Harbor Symp Quant Biol* **6**: 109–121. doi:10.1101/SQB.1938.006.01.013

9. Hall K. 2011. William Astbury and the biological significance of nucleic acids, 1938–1951. *Stud Hist Philos Bio Biomed Sci* **42**: 119–128. doi:10.1016/j.shpsc.2010.11.018

10. Hess EL. 1968. Origins of molecular biology. *Science* **168**: 664–669. doi:10.1126/science.168.3932.664

11. Although in fact it had been used by Warren Weaver in the Annual Report of the Rockefeller Foundation in 1938 to describe their funding for the "relatively new field, which may be called molecular biology." See Weaver W. 1970. Molecular biology: origin of the term. *Science* **170**: 581–582. doi:10.1126/science.170.3958.581-a

12. Chargaff E, Lipshitz R, Green C. 1952. Composition of the desoxypentose nucleic acids of four genera of sea-urchin. *J Biol Chem* **195**: 155–160.

13. Chargaff E. 1963. *Essays on nucleic acids*, p. 176. Elsevier, New York.

14. Watson JD, Crick FHC. 1953. Molecular structure of nucleic acids; a structure for deoxyribose nucleic acid. *Nature* **171**: 737–738. doi:10.1038/171737a0; Wilkins MHF, Stokes AR, Wilson HR. 1953. Molecular structure of deoxypentose nucleic acids. *Nature* **171**: 738–740. doi:10.1038/171738a0; Franklin RE, Gosling RG. 1953. Molecular configuration in sodium thymonucleate. *Nature* **171**: 740–741. doi:10.1038/171740a0

15. Watson JD, Crick FHC. 1953. General implications of the structure of deoxyribonucleic acid. *Nature* **171**: 964–967. doi:10.1038/171964b0

16. Olby R. 2009. *Francis Crick: hunter of life's secrets*, p. 256. Cold Spring Harbor Press, New York.

17. Schuck HN, et al. 1962. *Nobel: the man and his prizes*. Elsevier, Amsterdam.

18. Markel H. 2021. *The secret of life*, p. 420. Norton, New York.

19. Crick FHC. 1988. *What mad pursuit*, pp. 73–74. Basic Books, New York.

20. Meselson M, Stahl FW. 1958. The replication of DNA in *Escherichia coli*. *Proc Natl Acad Sci* **44**: 671–682. doi:10.1073/pnas.44.7.671

21. Because he had been diverted to work on mines during the Second World War.

22. Attributed to Rutherford in Bernal JD. 1939. *The social function of science*. p. 9. George Routledge, London. The exact occasion on which he may have said it has never been identified.

23. Quoted in Eve AS. 2013. *Rutherford. Being the life and letters of the Right Hon. Lord Rutherford, O.M.*, p. 183. Cambridge University Press, Cambridge. Letter from Rutherford to Hahn, December 22, 1908.

24. Kendrew JC. 1968. *Report of the working group on molecular biology*. HMSO, London.

25. Krebs H. 1969. *Biochemistry, "molecular biology" and the biological sciences*. Biochemical Society, London.

CHAPTER 14: DOGMA

1. Quoted in Judson H. 1979. *The eighth day of creation*, p. 337. Simon and Schuster, New York.

2. Crick FHC. 1958. On protein synthesis. *Symp Soc Exp Biol* **12:** 138–163.

3. An example from the early days of molecular biology is the comma-less code that Crick proposed in 1956, when he suggested that 20 triplet nucleotides could each be assigned to code for a unique amino acid in such a way that the other 44 triplets all had no meaning. This was a clever idea, but in fact the code is redundant and all the triplets have meaning.

4. Quoted in Crewdson J. 2002. *Science fictions*, p. 6. Little, Brown, New York.

5. Coffin JM. 2021. 50th anniversary of the discovery of reverse transcriptase. *Mol Biol Cell* **32:** 91–97. doi:10.1091/mbc.E20-09-0612

6. Coffin JM, Fan H. 2016. The discovery of reverse transcriptase. *Ann Rev Virol* **3:** 29–51. doi:10.1146/annurev-virology-110615-035556

7. Tooze J. 1970. Central Dogma reversed. *Nature* **226:** 1198–1199. doi:10.1038/2261198a0

8. Crick F. 1970. Central Dogma of molecular biology. *Nature* **227:** 561–563. doi:10.1038/227561a0

9. Discussion with David Baltimore, October 2021.

10. Tooze J. 1970. Après Temin, le Déluge. *Nature* **227:** 998. doi:10.1038/227998a0

11. Cairns J. 1978. *Cancer: science and society*. W.H. Freeman, San Francisco.

12. Bishop JM. 1980. The molecular biology of tumor viruses: a physician's guide. *N Engl J Med* **303:** 675–682. doi:10.1056/NEJM198009183031206

13. For the chronology of the isolation of cancer viruses, see Karpas A. 2004. Human retroviruses in leukaemia and AIDS: reflections on their discovery, biology and epidemiology. *Biol Rev Camb Philos Soc* **79:** 911–933. doi:10.1017/s1464793104006505

14. HIV was a compromise for the name resulting from disputes between Montagnier and Gallo about the nature of the virus.

15. Barré-Sinoussi F, et al. 1983. Isolation of a T-lymphotropic retrovirus from a patient at risk for acquired immune deficiency syndrome (AIDS). *Science* **220:** 868–871. doi:10.1126/science.6189183

16. Kuhn TS. 1962. *The structure of scientific revolutions*. University of Chicago Press, Chicago.

17. Kuhn's view about paradigm shifts is widely accepted, but his criticism of the view that science moves toward "objective truth" is widely rejected by scientists. I'm not sure this is entirely fair, as being a philosopher, he was concerned with what is meant by "truth." At all events, most scientists would claim to be moving toward objective truth.

18. Polyani M. 1946. *Science, faith, and society*. Oxford University Press, Oxford.

CHAPTER 15: DOCTRINES

1. In Conan Doyle A. 1890. *The Sign of Four*, Chapter 6. Spencer Blackett, London.

2. Popper often referred to a theory "proving its mettle."

3. Kraus A, et al. 2021. High-resolution structure and strain comparison of infectious mammalian prions. *Mol Cell* **81:** 4540–4551. doi:10.1016/j.molcel.2021.08.011

4. McKinley MP, Bolton DC, Prusiner SB. 1983. A protease-resistant protein is a structural component of the scrapie prion. *Cell* **35:** 57–62. doi:10.1016/0092-8674(83)90207-6

5. Wickner RB. 1994. [URE3] as an altered URE2 protein: evidence for a prion analog in *S. cerevisiae. Science* **264:** 566–569. doi:10.1126/science.7909170

6. Westaway D, et al. 1987. Distinct prion proteins in short and long scrapie incubation time period. *Cell* **51:** 651–662. doi:10.1016/0092-8674(87)90134-6

7. Discussion with Stanley Prusiner, November 2021.

8. Prusiner SB. 2014. *Madness and memory*. Yale University Press, New Haven, CT.

9. Cech TR, Zaug AJ, Grabowski PJ. 1981. *In vitro* splicing of the ribosomal RNA precursor of *Tetrahymena*: involvement of a guanosine nucleotide in the excision of the intervening sequence. *Cell* **27:** 487–496. doi:10.1016/0092-8674(81)90390-1

10. Guerrier-Takada C, McLain WH, Altman S. 1984. Cleavage of tRNA precursors by the RNA subunit of *E. coli* ribonuclease P (M1 RNA) is influenced by 3'-proximal CCA in the substrates. *Cell* **38:** 219–224. doi:10.1016/0092-8674(84)90543-9

11. Martick M, Scott WG. 2006. Tertiary contacts distant from the active site prime a ribozyme for catalysis. *Cell* **126:** 309–320. doi:10.1016/j.cell.2006.06.036

12. Kruger K, et al. 1982. Self-splicing RNA: autoexcision and autocyclization of the ribosomal RNA intervening sequence of *Tetrahymena. Cell* **31:** 147–157. doi:10.1016/0092-8674(82)90414-7

13. Discussion with Tom Cech, February 2022.

14. With typical understatement, he said, "[It] is not an enzyme but has some enzyme-like characteristics ... we call it a ribozyme, an RNA molecule that has the intrinsic ability to break and form covalent bonds." See note 12.

15. Yarus M. 2011. *Life from an RNA world: the ancestor within*. Harvard University Press, Cambridge, MA.

16. Eddington A. 1945. *New pathways in science. Messenger lectures*. Cambridge University Press, Cambridge, UK.

17. Chow LT, Gelinas RE, Broker TR, Roberts RJ. 1977. An amazing sequence arrangement at the 5' ends of adenovirus 2 messenger RNA. *Cell* **12:** 1–8. doi:10.1016/0092-8674(77)90180-5

18. Berget SM, Moore C, Sharp PA. 1977. Spliced segments at the 5' terminus of adenovirus 2 late mRNA. *Proc Natl Acad Sci* **74:** 3171–3175. doi:10.1073/pnas.74.8.3171

19. Discussion with Phil Sharp, October 2021.

20. Crick F. 1979. Split genes and RNA splicing. *Science* **204:** 264–271. doi:10.1126/science.373120

21. Stent GS. 1968. That was the molecular biology that was. *Science* **160:** 390–395. doi:10.1126/science.160.3826.390

22. This was connected with a generally lugubrious view of the role of science in progress, which was published as Stent GS. 1969. *The coming of the golden age: a view of the end of progress.* Natural History Press, New York.

23. See note 9 in Introduction, p. 13.

24. Quoted in Pais A. 1991. *Niels Bohr's times in physics, philosophy, and polity.* Oxford University Press, Oxford, UK.

Chapter 16: Mapping

1. At the Cold Spring Harbor meeting to celebrate the 100th anniversary of the birth of Francis Crick.

2. Painter TS. 1923. Studies in mammalian spermatogenesis. II. The spermatogenesis of man. *J Exp Zool* **37:** 291–336. doi:10.1002/jez.1400370303

3. Tjio JH, Levan A. 1956. The chromosome number in man. *Hereditas* **42:** 1–6. doi:10.1111/j.1601-5223.1956.tb03010.x

4. Justice MJ, et al. 1990. A genetic linkage map of mouse chromosome 10: localization of eighteen molecular markers using a single interspecific backcross. *Genetics* **125:** 855–866. doi:10.1093/genetics/125.4.855

5. The current status of the mouse genetic map is maintained in a database at www.informatics.jax.org.

6. Ingram VM. 1957. Gene mutations in human haemoglobin: the chemical difference between normal and sickle cell haemoglobin. *Nature* **180:** 326–328. doi:10.1038/180326a0

7. The adult organism has two copies of each chromosome, one of paternal origin and one of maternal origin. A sperm or egg has only one copy of each chromosome. If a chromosome passes into the sperm or egg without any change, it will be either paternal or maternal. *Recombination* describes what happens when there is an exchange to generate a chromosome that is partly paternal and partly maternal.

8. Dietrich WF, et al. 2005. Mapping the mouse genome: current status and future prospects. *Proc Natl Acad Sci* **92:** 10849–10853. doi:10.1073/pnas.92.24.10849

9. Botstein D, White RL, Skolnick M, Davis RW. 1980. Construction of a genetic linkage map in man using restriction fragment length polymorphisms. *Am J Hum Genet* **32:** 314–331.

10. Arber W. 1965. Host-controlled modification of bacteriophage. *Annu Rev Microbiol* **19:** 365–378. doi:10.1146/annurev.mi.19.100165.002053

11. Kelly TJ Jr, Smith HO. 1970. A restriction enzyme from *Hemophilus influenzae.* II. *J Mol Biol* **51:** 393–409. doi:10.1016/0022-2836(70)90150-6

12. Danna K, Nathans D. 1971. Specific cleavage of simian virus 40 DNA by restriction endonuclease of *Hemophilus influenzae. Proc Natl Acad Sci* **68:** 2913–2917. doi:10.1073/pnas.68.12.2913

13. http://rebase.neb.com/rebase/azlist.re2.html

14. RFLPs can be linked to known mutations. This was first done in humans for sickle cell anemia. Kan YW, Dozy AM. 1978. Polymorphism of DNA sequence adjacent to human β-globin structural gene: relationship to sickle mutation. *Proc Natl Acad Sci* **75:** 5631–5635. doi:10.1073/pnas.75.11.5631

15. Quoted in Zagorski N. 2006. Profile of Alec J. Jeffreys. *Proc Natl Acad Sci* **103:** 8918–8920. doi:10.1073/pnas.0603953103

16. Jeffreys AJ, Wilson V, Thein SL. 1985. Individual-specific fingerprints of human DNA. *Nature* **316:** 76–79. doi:10.1038/316076a0

17. Norman Anderson at Argonne National Laboratory wanted to catalog all the human proteins by using his technique of two-dimensional gel electrophoresis. Wade N. 1981. The complete index to man. *Science* **211:** 33–35. doi:10.1126/science.7444446

18. Sinsheimer RL. 1989. The Santa Cruz workshop. *Genomics* **5:** 954–956. doi: 10.1016/0888-7543(89)90142-0

19. Korenberg JR, et al. 1999. Mouse molecular cytogenetic resource: 157 BACs link the chromosomal and genetic maps. *Genome Res* **9:** 514–423. doi:10.1101/gr.9.5.514

20. Sawyer JR, Hozier JC. 1986. High resolution of mouse chromosomes: banding conservation between man and mouse. *Science* **232:** 1632–1634. doi:10.1126/science.3715469

21. Letter from Wally Gilbert to Robert Edgar (University of California), 1985. Cook-Deegan archive at Georgetown University (hdl.handle.net/10822/556964).

22. Cook-Deegan R. 1994. *The gene wars: science, politics, and the human genome.* Norton, New York.

23. From a tape made by Tom Caskey of comments at the discussion; transcript at the Cook-Deegan archive at Georgetown University (repository.library.georgetown.edu/handle/10822/559555).

24. Report on the human genome initiative. HERAC, 1987. Cook-Deegan archive at Georgetown University (hdl.handle.net/10822/556967).

25. See note 22, p. 144.

26. See note 22, p. 176.

27. For a report, see Roberts L. 1990. A meeting of the minds on the genome project? *Science* **250:** 756–757. doi:10.1126/science.2237425

28. Discussion with Tom Caskey, November 2021.

29. See note 27.

CHAPTER 17: GENOMES

1. Cold Spring Harbor Laboratory 1988 Annual Report, p. 5.

2. McKusick VA. 1990. *Mendelian inheritance in man*, pp. xi–xxix. Johns Hopkins University Press, Baltimore.

3. U.S. Department of Health and Human Services, U.S. Department of Energy: Understanding our Genetic Inheritance, The U.S. Human Genome Project: The First Five Years, Fiscal Years 1991–1995. http://www.ornl.gov/sci/techresources/Human_Genome/project/5yrplan/summary.shtml, 1990.

4. Data from the Human Gene Mapping Workshop in London in 1991 as summarized by McKusick VA. 1991. Genomic mapping and how it has progressed. *Hosp Pract* **26:** 74–90. doi:10.1080/21548331.1991.11705306

5. Angier N. 1990. Great 15-year project to decipher genes stirs opposition. *New York Times.* June 5, 1990.

6. Sanger F. 1988. Sequences, sequences, and sequences. *Annu Rev Biochem* **57:** 1–28. doi:10.1146/annurev.bi.57.070188.000245

7. Hunkapiller MW, Hood LE. 1983. Protein sequence analysis: automated micro-sequencing. *Science* **219:** 650–659. doi:10.1126/science.6687410

8. See note 6.

9. Maxam AM, Gilbert WA. 1977. A new method for sequencing DNA. *Proc Natl Acad Sci* **74:** 560–564. doi:10.1073/pnas.74.2.560

10. Amarasinghe SL, et al. 2020. Opportunities and challenges in long-read sequencing data analysis. *Genome Biol* **21:** 30. doi:10.1186/s13059-020-1935-5

11. Telenti A, et al. 2017. Deep sequencing of 10,000 human genomes. *Proc Natl Acad Sci* **113:** 11901–11906. doi:10.1073/pnas.1613365113

12. Shotgun cloning is faster, but is less certain than sequencing fragments that have already been ordered, because you have to find the overlaps after the sequencing has been done.

13. Fleishmann RD, et al. 1995. Whole-genome random sequencing and assembly of *Haemophilus influenzae* Rd. *Science* **269:** 496–512. doi:10.1126/science.7542800

14. Oral history recorded in 2006 at Cold Spring Harbor. http://library.cshl.edu/oralhistory/interview/genome-research/mechanics-hgp/venter-whole-genome-shotgun-sequencing/

15. Venter JC, et al. 2001. The sequence of the human genome. *Science* **291:** 1304–1351. doi:10.1126/science.1058040

16. International Human Genome Consortium. 2001. Initial sequencing and analysis of the human genome. *Nature* **409:** 860–921. doi:10.1038/35057062

17. For different estimates and methods of projecting the total number, see Antequera F, Bird A. 1993. Number of CpG islands and genes in human and mouse. *Proc Natl Acad Sci* **90:** 11995–11999. doi:10.1073/pnas.90.24.11995; Fields C, Adams MD, White O, Venter JC. 1994. How many genes in the human genome? *Nat Genet* **7:** 345–346. doi:10.1038/ng0794-345; Adams MD, et al. 1995. Initial assessment of human gene diversity and expression patterns based upon 83 million nucleotides of cDNA sequence. *Nature* **377:** 3–174.

18. Harrow J, et al. 2012. GENCODE: the reference human genome annotation for The ENCODE Project. *Genome Res* **22:** 1760–1774. doi:10.1101/gr.135350.111

19. After Pertea M, Salzberg SL. 2010. Between a chicken and a grape: estimating the number of human genes. *Genome Biol* **11:** 206–212. doi:10.1186/gb-2010-11-5-206

20. Sulston J, Ferry G. 2002. *The common thread*, p. 167. Joseph Henry Press, Washington, DC.

21. Based on comparison of protein sequences, see King MC, Wilson AC. 1975. Evolution at two levels in humans and chimpanzees. *Science* **188:** 107–116. doi:10.1126/science.1090005

22. The Chimpanzee Sequencing and Analysis Consortium. 2005. Initial sequence of the chimpanzee genome and comparison with the human genome. *Nature* **437**: 69–87. doi:10.1038/nature04072

23. Blattner FR, et al. 1997. The complete genome sequence of *Escherichia coli* K-12. *Science* **277**: 1453–1462. doi:10.1126/science.277.5331.1453

24. Goffeau A, et al. 1996. Life with 6000 genes. *Science* **274**: 546, 563–567. doi:10.1126/science.274.5287.546

25. The *C. elegans* Sequencing Consortium. 1998. Genome sequence of the nematode *C. elegans*: a platform for investigating biology. *Science* **282**: 2012–2018. doi:10.1126/science.282.5396.2012

26. The author list was originally provided as a link to a website. That site no longer exists, so the authors have been lost to posterity.

27. Adams MD, et al. 2000. The genome sequence of *D. melanogaster*. *Science* **287**: 2185–2195. doi:10.1126/science.287.5461.2185

28. The *Arabidopsis* Genome Initiative. 2000. Analysis of the genome sequence of the flowering plant *Arabidopsis thaliana*. *Nature* **408**: 796–815. doi:10.1038/35048692

29. See note 15.

30. See note 16.

31. Waterston RH, et al. 2002. Initial sequencing and comparative analysis of the mouse genome. *Nature* **420**: 520–562. doi:10.1038/nature01262

32. Church DM, et al. 2009. Lineage-specific biology revealed by a finished genome assembly of the mouse. *PLoS Biol* **5**: e1000112. doi:10.1371/journal.pbio.1000112

33. Based on the sequences of three different Neanderthal individuals. Green RE, et al. 2010. A draft sequence of the Neanderthal genome. *Science* **328**: 710–722. doi:10.1126/science.1188021

34. The 2% value is an average for comparison between Neanderthals and non-African humans. African humans have less Neanderthal DNA.

35. Genome sequences suggest that human and Neanderthal populations diverged between 270 and 440,000 years ago. Their common ancestor diverged from chimpanzees 5.6–8.3 million years ago.

36. Pääbo S. 2014. The human condition—a molecular approach. *Cell* **157**: 216–226. doi:10.1016/j.cell.2013.12.036

37. Discussion with Svante Pääbo, February 2022.

38. Zeberg H, et al. 2020. A Neanderthal sodium channel increases pain sensitivity in present-day humans. *Curr Biol* **30**: 3465–3469.e4. doi:10.1016/j.cub.2020.06.045

CHAPTER 18: EDITING

1. Discussion with Francisco Mojica, December 2021.

2. Ishino Y, Shinagawa H, Makino K, Amemura M, Nakata A. 1987. Nucleotide sequence of the *iap* gene, responsible for alkaline phosphatase isozyme conversion in *Escherichia coli*, and identification of the gene product. *J Bacteriol* **169**: 5429–5433. doi:10.1128/jb.169.12.5429-5433.1987

3. Mojica FJ, Ferrer C, Juez G, Rodríguez-Valera F. 1995. Long stretches of short tandem repeats are present in the largest replicons of the Archaea *Haloferax mediterranei* and *Haloferax volcanii* and could be involved in replicon partitioning. *Mol Microbiol* **17:** 85–93. doi:10.1111/j.1365-2958.1995.mmi_17010085.x

4. Mojica FJ, Díez-Villaseñor C, Soria E, Juez G. 2000. Biological significance of a family of regularly spaced repeats in the genomes of archaea, bacteria and mitochondria. *Mol Microbiol* **36:** 244–246. doi:10.1046/j.1365-2958.2000.01838.x

5. Jansen R, van Embden JD, Gaastra W, Schouls LM. 2002. Identification of genes that are associated with DNA repeats in prokaryotes. *Mol Microbiol* **43:** 1565–1575. doi:10.1046/j.1365-2958.2002.02839.x

6. Mojica scanned 4500 spacer sequences and found 88 that were present elsewhere, with two-thirds in viruses and one-third in other sources. Mojica FJ, et al. 2005. Intervening sequences of regularly spaced prokaryotic repeats derive from foreign genetic elements. *J Mol Evol* **60:** 174–182. doi:10.1007/s00239-004-0046-3

7. Pourcel C, Salvignol G, Vergnaud G. 2005. CRISPR elements in *Yersinia pestis* acquire new repeats by preferential uptake of bacteriophage DNA, and provide additional tools for evolutionary studies. *Microbiology* **151:** 653–663. doi:10.1099/mic.0.27437-0

8. The paper was rejected from *PNAS, Journal of Bacteriology, Nucleic Acids Research, and Genome Research.*

9. One *cas* gene was required to acquire resistance, but not to maintain it (suggesting that it might be concerned with inserting the spacer sequences); another was required to maintain resistance (suggesting that it is part of the mechanism that recognizes the sequences).

10. Barrangou R, et al. 2007. CRISPR provides acquired resistance against viruses in prokaryotes. *Science* **315:** 1709–1712. doi:10.1126/science.1138140

11. Marraffini LA, Sontheimer EJ. 2008. CRISPR interference limits horizontal gene transfer in staphylococci by targeting DNA. *Science* **322:** 1843–1845. doi:10.1126/science.1165771

12. Sontheimer E, Marraffini L. 2008. Target DNA interference with crRNA. U.S. Provisional Patent Application 61/009,317, filed September 23.

13. Brouns SJJ, et al. 2008. Small CRISPR RNAs guide antiviral defense in prokaryotes. *Science* **321:** 960–964. doi:10.1126/science.1159689

14. There are several different types of CRISPR systems, with the main distinction being whether a single enzyme (the product of *cas9*) cleaves the target DNA or whether this requires a complex of several proteins (called Cascade, in which the active subunit is the product of *cas3*). A type II system has four genes as described in the first reports, and a type I system has more genes to code for the components of Cascade. The genes have been renamed since the original discovery. The type II system has been used for most of the work to turn CRISPR into a gene-editing system.

15. Gasiunas G, Barrangou R, Horvath P, Šikšnys V. 2012. Cas9–crRNA ribonucleoprotein complex mediates specific DNA cleavage for adaptive immunity in bacteria. *Proc Natl Acad Sci* **109:** E2579–E2586. doi:10.1073/pnas.1208507109. Submitted to *Cell* April 6, rejected April 12; submitted to *PNAS* May 21, published online September 4.

16. Deltchev E, et al. 2011. CRISPR RNA maturation by trans-encoded small RNA and host factor RNase III. *Nature* **471**: 602–607. doi:10.1038/nature09886

17. Jinek M, et al. 2012. A programmable dual-RNA–guided DNA endonuclease in adaptive bacterial immunity. *Science* **337**: 816–821. doi:10.1126/science.1225829. Submitted June 8, accepted June 20, published online June 28.

18. Discussion with Francisco Mojica, December 2021.

19. Zhang's paper was submitted October 5, accepted December 12, published online January 3, and in print February 15. Church's paper was submitted October 26, and then accepted and published on the same dates as Zhang's.

20. Submitted December 15, 2012; accepted January 3, 2013; published online January 29, 2013.

21. Published as brief communications in *Nature Biotechnology*, so regarded more as technical results than major conceptual advances. Hwang WY, et al. 2013. Efficient genome editing in zebrafish using a CRISPR-Cas system. *Nat Biotechnol* **31**: 227–229. doi:10.1038/nbt.2501; Cho SW, Kim S, Kim JM, Kim JS. 2013. Targeted genome engineering in human cells with the Cas9 RNA-guided endonuclease. *Nat Biotechnol* **31**: 230–232. doi:10.1038/nbt.2507

22. Knot GJ, Doudna JA. 2018. CRISPR-Cas guides the future of genetic engineering. *Science* **361**: 866–869. doi:10.1126/science.aat5011

23. Isaacson W. 2021. *The codebreaker*, Ch. 28. Simon and Schuster, New York.

24. Lander ES. 2016. The heroes of CRISPR. *Cell* **164**: 18–28. doi:10.1016/j.cell.2015.12.041

CHAPTER 19: EPIGENETICS

1. Shakespeare W. *Julius Caesar*, Act I, Scene III, L. 140–141.

2. Heisenberg W. 1971. *Physics and beyond*, p. 206. Harper & Row, New York, as an account of a conversation Heisenberg had with Wolfgang Pauli and Niels Bohr in June 1952.

3. Heard E, Martienssen RA. 2014. Transgenerational epigenetic inheritance: myths and mechanisms. *Cell* **157**: 95–109. doi:10.1016/j.cell.2014.02.045

4. Dubois M. 2020. Epigenetics as an interdiscipline? Promises and fallacies of a biosocial research agenda. *Soc Sci Infor* **59**: 3–11. doi:10.1177/0539018420908233

5. A series of interactions leads to the assembly of a complex of proteins at the *FLC* gene that turns it off, so FLC protein is no longer made. The complex then keeps the gene in an inactive state.

6. This is known as the Polycomb repressor complex.

7. Discussion with Caroline Dean, November 2021.

8. Jacob F, Monod J. 1961. Genetic regulatory mechanisms in the synthesis of proteins. *J Mol Biol* **3**: 318–356. doi:10.1016/s0022-2836(61)80072-7

9. Waddington CH. 1968. *Towards a theoretical biology: the basic ideas of biology*, pp. 1–32. Edinburgh University Press, Edinburgh.

10. Holliday R. 1994. Epigenetics: an overview. *Dev Genet* **15**: 453–457. doi:10.1002/dvg.1020150602

11. Bird A. 2007. Perceptions of epigenetics. *Nature* **447:** 396–398. doi:10.1038/nature05913

12. One of the arguments against histone modification being more than part of the mechanism is that it's difficult to see how a specific target could be recognized. There is an argument that specificity is determined by RNAs, which can recognize the sequence of an RNA as it transcribed from the target gene, and then recruit the rest of the apparatus that modifies the histones. See Holoch D, Moazed D. 2015. RNA-mediated epigenetic regulation of gene expression. *Nat Rev Genet* **16:** 71–84. doi:10.1038/nrg3863

13. Discussion with Mark Ptashne, November 2021.

14. Mukherjee S. 2016. Same but different. *New Yorker*, May 2, 2016. The article was widely attacked by the scientific community for ascribing differences between twins to epigenetics. The text was revised before being included in Mukherjee's book, *The Gene* (see note 7 in Chapter 11).

15. Lyon MF. 1962. Sex chromatin and gene action in the mammalian X-chromosome. *Am J Hum Genet* **14:** 135–148.

16. Lyon MF. 1961. Gene action in the X-chromosome of the mouse (*Mus musculus* L.). *Nature* **190:** 372–373. doi:10.1038/190372a0

17. Discussion with Azim Surani, November 2021.

18. Rollin Hotchkiss at the Rockefeller University identified a modified base, which he called "epicytosine." Hotchkiss RD. 1948. The quantitative separation of purines, pyrimidines, and nucleosides by paper chromatography. *J Biol Chem* **175:** 315–332.

19. Bird A, Taggart M, Frommer M, Miller OJ, Macleod D. 1985. A fraction of the mouse genome that is derived from islands of nonmethylated, CpG-rich DNA. *Cell* **40:** 91–99. doi:10.1016/0092-8674(85)90312-5

20. Reik W, Collick A, Norris ML, Barton SC, Surani MA. 1987. Genomic imprinting determines methylation of parental alleles in transgenic mice. *Nature* **328:** 248–251. doi:10.1038/328248a0

21. Swain JL, Stewart TA, Leder P. 1987. Parental legacy determines methylation and expression of an autosomal transgene: a molecular mechanism for parental imprinting. *Cell* **50:** 719–727. doi:10.1016/0092-8674(87)90330-8

22. Leonhardt H, Page AW, Weier HU, Bestor TH. 1992. A targeting sequence directs DNA methyltransferase to sites of DNA replication in mammalian nuclei. *Cell* **71:** 865–873. doi:10.1016/0092-8674(92)90561-p

23. Surani MA, Reik W, Norris ML, Barton SC. 1986. Influence of germline modifications of homologous chromosomes on mouse development. *J Embryol Exp Morphol* **97 Suppl:** 123–136.

24. Epigenetic changes resulting from the environment are sometimes attributed as the cause of differences develop between identical twins as they age, especially when they are correlated with changes in DNA methylation patterns. But there is no case yet in which mutations or other random changes have been excluded as the cause.

25. Although it must have a different basis in flies (e.g., *Drosophila*) and worms (e.g., *C. elegans*), in which there is no methylation.

26. McClintock B. 1950. The origin and behavior of mutable loci in maize. *Proc Natl Acad Sci* **36:** 344–355. doi:10.1073/pnas.36.6.344. See Fedoroff N. 2012. McClintock's challenge in the 21st century. *Proc Natl Acad Sci* **109:** 20200–20203. doi:10.1073/pnas.1215482109

27. McClintock B. 1961. Some parallels between gene controls in maize and in bacteria. *Am Nat* **95:** 265–277. doi:10.1086/282188

28. Cropley JE, Suter CM, Beckman KB, Martin DI. 2006. Germ-line epigenetic modification of the murine *A^vy* allele by nutritional supplementation. *Proc Natl Acad Sci* **103:** 17308–17312. doi:10.1073/pnas.0607090103

29. The CpG sequences are at a site that turns off the *agouti* gene. The site is a transposon, the same sort of element that has been associated with epigenetic effects in maize.

30. Lumey LH, et al. 1993. The Dutch famine birth cohort study: design, validation of exposure, and selected characteristics of subjects after 43 years follow-up. *Paediatr Perinat Epidemiol* **7:** 354–367. doi:10.1111/j.1365-3016.1993.tb00415.x

31. Heijmans BT, et al. 2008. Persistent epigenetic differences associated with prenatal exposure to famine in humans. *Proc Natl Acad Sci* **105:** 17046–17049. doi:10.1073/pnas.0806560105

32. McClintock showed that elements in maize can cycle between active and inactive phases that are inherited; we now know that these are examples of transposons (elements that can move within the genome). See Jones RN. 2005. McClintock's controlling elements: the full story. *Cytogenet Genome Res* **109:** 90–103. doi:10.1159/000082387; Fedoroff NV. 2012. McClintock's challenge in the 21st century. *Proc Natl Acad Sci* **109:** 20200–20203. doi:10.1073/pnas.1215482109

33. Paramutation is a phenomenon in maize discovered by Alexander Brink in the 1960s, in which an interaction between the two copies of a gene results in an inherited change in the state of gene expression. See Brink RA. 1960. Paramutation and chromosome organization. *Q Rev Biol* **35:** 120–137. doi:10.1086/403016

34. Epigenetic inheritance in *C. elegans* may be due to small RNAs; for example, see Ashe A, et al. 2012. piRNAs can trigger a multigenerational epigenetic memory in the germline of *C. elegans*. *Cell* **150:** 88–99. doi:10.1016/j.cell.2012.06.018

35. Daxinger L, Whitelaw E. 2012. Understanding transgenerational epigenetic inheritance via the gametes in mammals. *Nat Rev Genet* **13:** 153–162. doi:10.1038/nrg3188

36. This is the case with *agouti*, in which the variable expression of yellow is partly inherited by progeny from their mothers. It appears that the state of expression of the gene is always reset in sperm but only partially reset in eggs. Morgan HD, Sutherland HG, Martin DI, Whitelaw E. 1999. Epigenetic inheritance at the agouti locus in the mouse. *Nat Genet* **23:** 314–318. doi:10.1038/15490

CHAPTER 20: STEM CELLS

1. "Every Sperm Is Sacred" is a musical sketch from *Morty Python's The Meaning of Life*, 1983.

2. C.P. Snow wrote a series of novels exposing the machinations of politicking at Oxford and Cambridge in the 1950s. (He was already well-known for his books when he gave his lecture on *The Two Cultures:* see note 1 in Introduction.)

3. Gurdon JB, Laskey RA, Reeves OR. 1975. The developmental capacity of nuclei transplanted from keratinized skin cells of adult frogs. *J Embryol Exp Morph* **34:** 93–112.

4. See note 17 in Chapter 11.

5. Mostly because of abnormal fetal development.

6. Hochedlinger K, Jaenisch R. 2003. Nuclear transplantation, embryonic stem cells, and the potential for cell therapy. *N Engl J Med* **349:** 275–286. doi: 10.1056/NEJMra035397

7. Thomson JA, et al. 1998. Embryonic stem cell lines derived from human blastocysts. *Science* **282:** 1145–1147. doi: 10.1126/science.282.5391.1145

8. Takahashi K, et al. 2007. Induction of pluripotent stem cells from adult human fibroblasts by defined factors. *Cell* **131:** 861–872. doi: 10.1016/j.cell.2007.11.019

9. Takebe T, et al. 2013. Vascularized and functional human liver from an iPSC-derived organ bud transplant. *Nature* **499:** 481–484. doi: 10.1038/nature12271

10. Khampang S, et al. 2021. Blastocyst development after fertilization with in vitro spermatids derived from nonhuman primate embryonic stem cells. *Fl Sl Sci* **2:** 365–375. doi: 10.1016/j.xfss.2021.09.001

EPILOGUE: REDUCTIONISM

1. Crick F. 1966. *Of molecules and men*, p. 10. University of Washington Press, Seattle.

2. The origin of this saying appears to lie with Kluyver in 1926: "From the elephant to butyric acid bacterium—it is all the same!" (1959. *Albert Jan Kluyver: his life and work* [ed. Kamp AF, et al.]. North-Holland, Amsterdam.) It has been attributed to Jacques Monod, and dated to 1954, for decades, but I have been unable to trace the occasion on which he said it.

3. Crick FH, Orgel LE. 1973. Directed panspermia. *Icarus* **19:** 341–346.

4. Lewin B. 1998. The mystique of epigenetics. *Cell* **93:** 301–303. doi:10.1016/s0092-8674(00)81154-x

5. This partly depends on how reductionism is defined. See for example Weinberg S. 2003. *Facing up: science and its cultural adversaries*, p. 111. Harvard University Press, Cambridge, MA.

6. Mayr E. 1985. *The growth of biological thought. Diversity, evolution, and inheritance.* Harvard University Press, Cambridge, MA.

7. Weinberg S. 1987. Newtonianism, reductionism and the art of congressional testimony. *Nature* **330:** 433–437. doi:10.1038/330433a0

8. Anderson PW. 1972. More is different. *Science* **177:** 393–396. doi:10.1126/science.177.4047.393

9. Horgan J. 2015. *The end of science*. Basic Books, New York.

10. Planck M. 1933. *Where is science going?* Allen & Unwin.

11. Crick FHC. 1994. *The astonishing hypothesis*. Scribner, New York.

ILLUSTRATION SOURCES

CHAPTER 1

Page 20: Courtesy DeepMind.

CHAPTER 6

Page 85: Courtesy of Richard Myers.

CHAPTER 7

Page 89: Reprinted with permission from Springer Nature.

Page 94: https://commons.wikimedia.org/wiki/File:Institute_for_Scientific_Information_(ISI).jpg

CHAPTER 11

Page 155: Courtesy Zeiss.

CHAPTER 12

Page 165: Courtesy American Philosophical Society, Curt Stern Papers.

Page 169: https://upload.wikimedia.org/wikipedia/commons/8/86/T4_phage.jpg

CHAPTER 13

Page 173: https://commons.wikimedia.org/wiki/File:The_Cavendish_Laboratory_-_geograph.org.uk_-_631839.jpg

Page 174: https://en.wikipedia.org/wiki/Photo_51#/media/File:Photo_51_x-ray_diffraction_image.jpg

Page 175: A. Barrington Brown/Science Photo Library.

Page 176: Reprinted from *Nature* V.171, 1953, 737–738, with permission from Springer Nature.

CHAPTER 14

Page 184: Courtesy Wellcome Library, London.

Page 188: Courtesy CSHL.

Page 190: Courtesy NIAID.

CHAPTER 15

Page 195: Courtesy Allison Kraus, Austin Athman, Efrosini Artikis, Byron Caughey (NIAID).

Page 203: Reprinted with permission from Elsevier.

CHAPTER 16

Page 214: Reprinted from *Nature* V.316, 1985, 76–79, with permission from Springer Nature.

Page 217: Courtesy National Human Genome Research Institute.

CHAPTER 17

Page 224: Reprinted from Maxam AM, Gilbert W. 1977. *Proc Nat Acad Sci* **74:** 560–564, with permission from W. Gilbert.

Page 225: Courtesy National Human Genome Research Institute.

Page 226: Photo © Sam Ogden.

CHAPTER 19

Page 246: https://plantlet.org

Page 254: Reprinted with permission from *Proc Nat Acad Sci* **103:** 17308–17312, © 2006 National Academy of Sciences, U.S.A. Courtesy David I.K. Martin.

CHAPTER 20

Page 259: https://en.wikipedia.org/wiki/African_clawed_frog/Poleta 33

Page 260: Courtesy Roslin Institute, University of Edinburgh.

INDEX

Note: Page numbers in *italics* denote figures, tables, or photographs on the corresponding page.